T0178113

Rural Development Theory and Practice

Routledge Studies in Development and Society

Rural Development Theory and Practice

Ruth McAreavey

Routledge
Taylor & Francis Group
New York   London

First published 2009
by Routledge
711 Third Ave, New York, NY 10017

Simultaneously published in the UK
by Routledge
2 Park Square, Milton Park, Abingdon, Oxon OX14 4RN

Routledge is an imprint of the Taylor & Francis Group, an informa business

First published in paperback 2012

Typeset in Sabon by IBT Global.

Library of Congress Cataloging-in-Publication Data
McAreavey, Ruth.
  Rural development theory and practice / by Ruth McAreavey.
    p. cm. — (Routledge studies in development and society ; 19)
    Includes bibliographical references and index.
  1. Rural development. I. Title.
    HN49.C6M234 2009
    307.1'412—dc22
    2008047609

ISBN10: 0-415-95764-8 (hbk)
ISBN10: 0-203-87812-4 (ebk)

ISBN13: 978-0-415-95764-9 (hbk)
ISBN13: 978-0-203-87812-5 (ebk)
ISBN13: 978-0-415-65156-1 (pbk)

Contents

Acknowledgments

I am indebted to many people and I would not be able to mention them all here. That is not to belittle each of their input: I am extremely grateful to my family along with the colleagues, teachers and mentors that I have had throughout my career. I would nonetheless like to mention just a few by name.

Rachel Woodward and Philip Lowe both contributed to the development of my ideas around notions of rural development. Trevor Newsom helped these ideas come to fruition. Meanwhile Sally Shortall has been entirely inspiring since the first day I met her. It was Sally who gave me the idea to write this book and I am grateful for her ceaseless encouragement, advice and support.

Finally, just in case I was in danger of retreating into an esoteric world of rural development, power and micro-politics, my young sons Isaac and Luca have been there to keep me grounded. This book has been written between tantrums, teething and general toddler turmoil. So my thanks to Steve for continued listening and good humour.

1 Introduction and Overview

This book is a study of the policy and practice of rural development. It is not simply about rural development policy, nor does it solely focus on resulting activity. Policy and practice are linked through processes of governance, and it is through this lens that rural development is examined. Using examples from case study material, the book addresses a knowledge deficit that exists between policy and practice while also strengthening our understanding of the governance of rural areas. In so doing it is hoped that it will contribute to existing debates within rural development and that it will inform policy development.

Some of the incentive for writing this book derives from my personal experience as a rural development practitioner. In this sense it is a practically motivated study in that it aims to consider the policy and practice of rural development and the links, or lack thereof, between the two. In this manner it is hoped to contribute to our knowledge base. Moreover there are political and intellectual rationales for pursuing this line of investigation. Having become aware of divergences between project and programme activity whilst employed as a rural development professional on the ground, I was concerned that while being acknowledged informally by both practitioners and policymakers, nothing more forceful was being done about it. The gap in the knowledge and action aroused my curiosity. Often objectives of grandiose policies appeared to be esoteric and far removed from the values, beliefs and indeed material resources of actors in the field. I believe that better knowledge ought to lead to more effective policy and practice. This book aims to raise policy and academic debate in this regard and to highlight areas for future research. Here I simply reiterate its over-arching objective: it seeks to critically analyse the policy and practice of rural development. The study is undertaken in the context of the new rural governance and with attention to the role of power vis-à-vis social relations and structures. It is on this basis that we can proceed with an analysis of rural development that seeks to advance our knowledge of the links between knowing and doing. Therefore the study has practical, political and intellectual purposes. It will contribute to future research in this area, it will enhance the knowledge base of practitioners and policymakers and it will place centre stage the dynamics of power relations within the sector.

POLICY AND PRACTICE

This book considers the interplay between the policy and practice of rural development and regeneration. It claims that due to the constraints of, and relations between, the institutions, mechanisms and practice of rural development, actors do not have equal access to power and power is exercised unevenly by these actors. As a result the extent and effectiveness of rural development itself is limited in terms of who is involved and in relation to the ultimate impact of rural policy. The study will consider power relations within rural development and the relations and interplay between policy and practice. It aims to draw attention to the role of power within rural development.

Power relations extend beyond practitioners to other stakeholders such as policymakers. Ultimately these stakeholders do not necessarily have equal power status. This study will investigate where power is derived and exercised in the complex framework. It will reveal the import of these dynamics by showing the effect on policy and practice. It is argued that disparities in power relations result in a gap between the policy and practice of rural development. This is manifest in terms of what is claimed to be achievable, what actually happens and eventually what is possible in the future.

But the focus of the book transcends power; it is fundamentally about the analysis of the policy and practice of rural development in the context of the new rural governance. How are we to understand and make a connection between policy and practice? Like two sides of a coin, one relates to the other. This is in complete contrast to the writings of Auguste Comte who believed that social reform was a theoretical, cerebral enterprise, set apart from the practical activity of the individual on the street. Coining the term *positivism* he believed that the real world exists separately to peoples' perceptions or beliefs of it, it is not constructed by people. Pure theoretical projects could be devised that were based on general social laws that emanated from the 'science social'. Implementation he believed was an insignificant, practical detail that followed from intellectual efforts. Traditional, rational policy analysis is rooted in Comte's beliefs, whereby political decision making and politics are segregated as much as possible. And so, as Hajer and Wagenaar (2003) argue, rational policy analysis has led to a gap between theoretical rationality of policy and the practical rationality of the practitioner. It is precisely this gap which is of interest to our analysis. Turning practice on its head, Hajer and Wagenaar maintain that it is a theoretical concept, attempting to connect knowing and doing and so 'acting is the high road to knowing' (2003:20). They reveal its complex nature, showing how it is not just a mere translation of action but it is about the actor and his or her beliefs and values, it denotes the interdependence of social, individual and material concerns and it presupposes the social. These are matters to which we will return throughout the book.

As we settle into the twenty-first century, the multi-faceted nature of the challenges facing rural areas is becoming increasingly evident. Whereas in the past rural equated to agriculture, this is no longer the case for communities across the globe, be they in rural China or in North America. The complexity of the issues means that one-dimensional sectoral policies can no longer adequately address the needs of rural areas. Using examples from across the globe, Chapter 2 charts the progress of rural and agricultural policy. Integrated rural development, multi-functionality and bottom-up approaches are some of the concepts that are scrutinised. Overall this chapter provides context for the emergence of the new rural governance.

There has been a tendency to embrace the new vocabulary of governance without much rigorous analysis and empirical investigation (Beck, 1999). This book uses empirical case study material based on research conducted in rural England. Chapter 3 presents a brief socio-economic description of each of the communities and an overview of the type of regeneration activity that emerged from the case studies. It is clear that while the research relates to specific areas, the issues are relevant beyond those areas, having resonance with many western rural communities. This chapter describes the framework within which the research was conducted, the significance of this becoming clear as the book progresses and the nature of the issues emerging within regeneration practice is revealed. Where appropriate pseudonyms and codes have been applied throughout the book to conceal the identity of specific individuals. The actual research process itself is significant and represents a novel approach to ethnographic study (see McAreavey, 2008 for a full analysis).

The political nature of the rural development domain is demonstrated by the fact that establishing the legitimacy of the project was linked to power and politics (Coghlan and Brannick, 2001) as was the management of my relations with superiors, peers and colleagues (Kotter, 1985). These matters will unfold given the centrality of issues of power to this book. Power specifically is addressed in the chapter following which is necessarily theoretical. It explores in some depth the notion of power and the meaning that is given to it throughout this book. The personal nature of power is highlighted, albeit within a broader context of social structures. Critical to this analysis is Lukes' three faces of power, the nature of 'real' interests and the ability of individuals to change the structures within which they exist. This chapter illustrates how a close analysis of processes within rural development is necessary to understand power relations. That analysis is the objective of the proceeding chapters.

It has been noted that 'project controversies are problems based on personality conflicts, miscommunication, and misinformation rather than more fundamental matters of value or principle' (Lowry et al., 1997:181).

Chapters 5 and 6 focus on these 'intangibles' that bind groups together. Categorising these issues as micro-politics, Chapter 5 reveals how it relates to trust; power; and personal attributes such as perceptions and motivations. It shows the similarities and differences between social capital and micro-political processes. As micro-political processes underpin and have commonalities with the characteristics of social capital, a better understanding of micro-politics will contribute to our knowledge of this versatile concept. But further, it becomes clear that understanding micro-politics is pivotal to gaining a deeper understanding of the interests of actors in the rural development process. As we seek to comprehend power relations among agents, this conceptual framework becomes an essential tool. The import of micro-politics is demonstrated in Chapter 6 using detailed examples from the ethnographic study. Different aspects of group dynamics are critically analysed, along with their relevance to micro-politics. Micro-political processes are often unintended consequences of community action, and they are difficult to measure and so are sidelined within the development process. Nonetheless the importance of micro-politics cannot be ignored and the chapter shows how positive micro-politics have intrinsic value while also contributing to more tangible policy objectives.

Chapters 7 and 8 take up the notion of participation. They chart the meaning assigned to this concept in an era of globalisation and decentralisation. The seminal work of Arnstein during the late 1960s provides a useful starting point for analysis. Not only is participation shown to be ubiquitous but it is also an all encompassing label. Drawing extensively from the empirical data, these chapters use 'thick description' to enlighten our comprehension of participation policies and practice. The analysis seeks to understand how individuals operating within regeneration give meaning to participation before scrutinising in some depth what it means to participate in rural development activities. The enquiry uses the preceding discussions on power and micro-politics to consider power relations among agents. It views participatory practices from top-down and bottom-up perspectives in an attempt to understand the role of the state in this process. The chapters indicate that participating in regeneration activities is not what it seems; power relations are asymmetrical, and a regeneration power elite exists. Questions must therefore be asked about where benefits flow and how regeneration activities are framed.

Finally, a note on jargon. Where possible I have attempted to minimise the number of acronyms used and to simplify descriptions of institutional structure. For those familiar with the multiplicity of organisations operating within rural governance, you will understand that this is not an easy task. Doubtless the specifics of the terms *rural regeneration* and *rural development* could be debated at length; however throughout this book, in a bid to minimise confusion, the labels are used interchangeably. Any other particulars concerning terminology are addressed in relevant chapters.

2 Rural Areas in the 21st Century

Rural areas have changed dramatically over the past fifty years. Challenges have shifted and new opportunities have emerged. The role of the family farm has been eroded. Indeed the role of agriculture in the rural economy has diminished with less than 10% of the rural workforce in Organisation for Economic Co-operation and Development (OECD) countries employed in agriculture. Meanwhile in the European Union (EU) while 96% of land is in agriculture, only 13% of rural employment is in agriculture (OECD, 2006). People believe there should be a greater good emerging from agriculture than simply food production; they want to gain from the landscape aesthetically and they also want to be able to use rural areas as an amenity. In other words rural resources are seen as being multi-functional as they give private and public benefits. Meanwhile as the effects of 'Agflation', that is, the rising cost of agricultural production due to the global credit crunch and the rising cost of fuel, are felt by all nations, the consumer is increasingly demanding cheaper and higher quality food. Food security arguments have re-emerged in recent years within international trade negotiations as a means of justifying subsidies. Issues of public health and animal welfare have risen to prominence, for example the outbreak of foot-and-mouth disease in the United Kingdom (UK) in 2001 and the avian influenza (H5N1) outbreak in Southeast Asia in mid-2003 (which since spread in Asia and to Europe). A sneeze in one country very quickly reverberates around the globe.

Rural areas have been opened up, agriculture is no longer able to exist by its own standards and a multitude of stakeholders have identified themselves as having an interest in rural matters. National boundaries no longer denote legislative limits. By necessity the focus of rural policy has altered, and new approaches to rural development attempt to take into account the differential nature of rural areas in terms of assets and needs.

This chapter provides context for the emergence of the new rural governance. It charts the progress of rural and agricultural policy, using examples from across the globe. Integrated rural development, multi-functionality and bottom-up approaches are some of the concepts that are unpacked. The significance of public sector reform in the era of globalisation and decentralisation is considered to illustrate the importance of participation in the 21st century.

DYNAMICS OF RURAL AREAS

Population mobility is evident in rural areas. While some outlying areas will always struggle to retain population, on the whole the mobility of the population is striking. There are currently about 40 million foreign nationals (8.6%) in the EU25 countries (Katseli et al., 2006). Meanwhile the United States (US) Census Bureau's 2006 American Community Survey reported 37,547,789 foreign born individuals in the United States, which represents 12.5% of the total US population (Terrazas et al., 2007). Rural and urban areas alike have been recipients of foreign nationals, and countries with little past experience of immigration have become destination areas for migrants in the 21st century (Grillo, 2001; Penninx et al., 2008; Pollard et al., 2008). Nonetheless people are still moving to urban areas, rural areas are ageing and educational attainment is lower as are levels of public service (OECD, 2006).

Not only is the profile of rural areas changing, but the perception of the role of rural areas held by the masses has shifted. A significant problem concerns the ways in which modes of land use are affecting water quality and broader ecosystems (United Nations Environment Programme [UNEP], 2007). Whereas once the farming community was at liberty to farm land as they saw fit, today the subjects of climate change, the environment and sustainable development more generally are prominent in social discourse. Setting aside the problems with these contentious issues, the consequence is that land is viewed as a valuable natural resource and so is subject to control through regulation and management. Hence we see the rise of standards for water, soil and air quality. The volume of legislation emanating from national and international bodies with the aim of ameliorating the perceived problems resulting from land use is proof of the level of societal concerns. For example in Europe this is especially evident in the Water Framework Directive (WFD), the Strategic Environmental Assessment Directive, the Nitrates Directive and cross-compliance regulations of Common Agricultural Policy (CAP) reforms.

Different approaches can be taken to achieving targets; enforcement and/or incentivising schemes. For example in New Zealand the polluter-pays principle is used without the use of subsidies so that the polluter bears the cost of ensuring the environment is in an acceptable state. Resource management and landcare programmes are also used to allow local authorities, landowners and community groups to take a proactive role, and activities include treeplanting, monitoring biodiversity and providing information to local authorities (http://www.maf.govt.nz/mafnet/rural-nz/sustainable-resource-use/resource-management/agrienvironmental-programmes/httoc.htm, last accessed 15.08.08). Similarly in Saskatchewan, Canada,

Environmental Farm Plans provide a voluntary means for producers to identify management practices to reduce environmental risk on their farms. This is complemented by the Agri-Environmental Group Planning project whereby either sectoral groups or territorial groups identify an agri-environmental priority issue that may be addressed through financial support from the Canada-Saskatchewan Farm Stewardship Program. Environmental quality may be achieved through statutory legislation or through positive management plans. The latter may be targeted at individual landowners or at stakeholder groups, typically using structures of governance.

Simultaneously, modernisation and reform of the public sector are increasingly evidenced across the globe (OECD, 2005). National governments struggle with a range of issues that have come to the fore within 21st century industrialised nations, such as managing ever more scarce budgets; delivering services and basic levels of welfare to ageing populations; integrating increasingly diverse foreign born nationals and ensuring adequate mechanisms for the active engagement and participation of citizens. These modernisation projects represent substantial reform proposals as was evidenced recently in Ireland (OECD, 2008). The intricacy of the implementation of such programmes can be seen from the modernisation agenda that arose from the decentralisation of responsibilities for policy making and delivery in the UK. This was part of a wider programme of devolution and constitutional change that was introduced by New Labour under Tony Blair's leadership in 1997. It has been gathering pace since the late 1990s, and English devolution has been about building up regional administrative capacity to take account of territorial diversity, rather than elected regional government (Pearce et al., 2005). As rural policy is implemented locally it inevitably becomes engaged with these systems of governance so that a myriad of structures is cultivated.

Across the globe there is evidence of a sea change within structures of government. Nation-states increasingly participate at a global level through organisations and networks such as the World Trade Organisation or the G8 group. Intermediate associations remain important mechanisms for agreements and co-operation as demonstrated through groups such as the EU or the Southern African Development Community. Decentralisation of government structures has occurred at the regional level, with evidence of reform of local government across Europe and of emerging increased responsibilities for local areas shifting from central to local government. Consequently a complex of multi-scalar government has emerged within many nation-states. This is evident in Northern Ireland where the current implementation of the Review of Public Administration will result in a range of increased powers for local councils (Northern Ireland Executive, 2006).

And so there is evidence of the ascent of concepts including multi-functionality, stewardship, public goods, governance and devolution within rural policy rhetoric (OECD, 2001a, 2001b, 2006). These themes are embedded within European rural policy. From 1988 it began to move away from a prescriptive top-down, sectoral approach to incorporate a stronger territorial, spatial dimension that acknowledges the need to integrate social, economic and environmental issues (Commission of the European Communities [CEC], 1988). The associated and ongoing reform of the CAP resulted in the creation of new institutional apparatus, including the forging of new relationships between rural actors and the creation of a plethora of partnerships (Ward and McNicholas, 1998). This approach retains currency through Europe's Agenda 2000 reform (CEC, 1998) and is manifest within the second pillar of the CAP with the establishment of the Rural Development Regulation within which the Liason Entre Actions de Dévelopment de l'Economie Rurale (LEADER) approach has been mainstreamed (CEC, 2005). Current rural policy reveals a shift from purely agricultural matters to encompass a territorial and integrated approach that takes account of economic, social and environmental needs of particular rural areas (CEC, 1998; Shortall and Shucksmith, 2001; Shortall, 2004; Ward and Lowe, 2004; CEC, 2005). We will return to the example of Europe later in the analysis.

RURAL REGENERATION AND RURAL DEVELOPMENT

As a result of these pressures and transformations, the institutional infrastructure for rural regeneration and development has undergone many changes. The effect has been the emergence of multi-scalar governance whereby there is an increase in the volume of partnerships denoting a new relationship between the state and its citizens. Moreover many of the structures are superimposed on one another to provide a somewhat chaotic governance environment for rural development and regeneration activities.

Within this broader global context this book focuses on the emergence of decentralised arrangements to support the delivery of rural development and regeneration initiatives at a local level in the pursuit of endogenous development strategies. They are premised on the notion that such development approaches will be more successful because they start from the local resource base and also involve local participation in the design and implementation of development action (Ray, 1999a). In practice the concepts of rural development and rural regeneration intertwine and overlap, even though they have a slightly different heritage. Broadly rural development policy has emerged from agricultural reform while regeneration has surfaced as part of a global agenda of neo-liberalisation (see Chapter 1 for further discussion). Two key influencing factors are relevant; one is decentralisation and the other is the reform of agricultural policy. These merit closer attention and are considered in turn.

Decentralisation and Regionalisation

The role of the nation-state has been re-positioned as a result of globalisation and post-Fordism resulting in a new political arrangement comprising multi-level metagovernance (Jessop, 2005). Governance marks a departure from conventional styles of government with the erosion of traditional boundaries and reliance and involvement of partners beyond government (Stoker, 1998; Jessop, 2002; Goodwin, 2003). Decentralisation and the reformation of political structures bring the concept of participation centre stage. It enables citizens to more actively participate in structures of 'governance', that is, an institutional framework broader than government, based on the idea of partnership, devolving power, and including the community, public and private sectors (Jessop, 1990; Tendler, 1997; Jessop, 2002). A proliferation of partnerships between public and civil society sectors stemming from these new governance practices has been evident across the globe (Rhodes, 1997; Goodwin, 1998; Lowndes and Skelcher, 1998; Stoker, 1998; Cheverett, 1999; Jones and Little, 2000; Pierre, 2000; Edwards et al., 2001; Jessop, 2002; Gaventa, 2004; World Bank, 2004; Bryden, 2005).

The result of the trend in regional policy to decentralise is far reaching, leading to many challenges. It has necessitated a paradigm shift from a top-down approach to one which relies more on a bottom-up, integrative approach, involving many different partners: the state no longer assumes sole responsibility for governing. This sea change to the implementation of policy relies on local assets and knowledge (OECD, 2006). It calls for a collective/negotiated approach which revolves around power relations as actors seek to influence the actions of others in order to pursue their own agenda. This has not been without criticism and the literature would suggest that the capacity for communities to exercise genuine power is questionable (Hastings, 1996; Bochel, 2006; Gilchrist, 2006; Taylor, 2007). This is something to which we will return throughout the chapter and indeed the book.

New opportunities are created as citizens are increasingly encouraged to participate in structures of governance across the globe. But the effects of regionalisation and globalisation have placed pressure on traditional hierarchical administrative structures (OECD, 2006). Currently many systems of local government in Europe are undergoing substantial reform with emerging models designed to engage more effectively with users and stakeholders in a particular area, while also providing a mechanism whereby legislation may be successfully implemented. In Northern Ireland the Review of Public Administration aims to strengthen the role of local government so that

> under the new system councils will have responsibility for a wide range of functions and a strong power to influence a great many more. This will enable them to respond flexibly to local needs and make a real difference to people's lives . . . And through community planning the opportunity exists to promote good relations, address poverty and environmental issues, and develop normal civic society . . . Councils

will not necessarily directly deliver all the services for which they are responsible. They will be encouraged to develop partnership arrangements with the voluntary and community sectors, and the private sector in developing and commissioning services. (Northern Ireland Executive, 2006:7, 8)

In other words, at the core of the new council design is reliance on a locally based partnership approach.

Citizen Empowerment and Democracy

Callanan (2005) shows how a variety of different participative mechanisms are used by governments across Europe to facilitate participatory democracy. They are positioned at different points along a spectrum defining the relationship between the state and its citizens and stakeholders and the accompanying power relations. Participation, he reveals, may be valued for the process itself, or for the results that it produces. Back in the 1960s Arnstein categorised participation using the principle of a hierarchical ladder. Participation is closely connected to power as she claims 'citizen participation is a categorical term for citizen power' (Arnstein, 1969:216). The model was particularly insightful and radical; the ladder ranges from manipulation and therapy at the bottom where non-participation occurs to the top where full participation and citizen control predominate. It advocates a goal of full participation and with this, citizen empowerment gained from citizen control. Unless fully in control, according to Arnstein's theory, citizens are powerless, they do not have the ability to exert influence or to make changes or to resist change (Locke, 1979; Lukes, 2005). This does not leave any room for compromise and collaboration as advocated by current approaches to governance (OECD, 2006).

Arnstein's model advocates a goal of full participation and, with this, empowerment of the citizen through complete control. These are not necessarily the objectives of participatory techniques in rural development and regeneration today: while government policy often continues to encourage full participation, citizen control is no longer an objective. Instead, contemporary policy is about making 'government more responsive by enabling citizens to participate in decision making' (OECD, 2005:3) so that local government aims to use its powers to enable rather than control (Taylor, 2000). International development programmes typically encourage participation 'to improve service delivery, empower the poor and enhance participation of local communities in their own development . . . and to enhance local governance and local institution building' (Babajanian, 2005:449). Lowndes et al. (2001a) considered public participation as attempts by local government in England to encourage participation in local affairs beyond the traditional processes of political engagement, that is, voting and party membership. They found that the importance of gaining citizens' views was seen to be useful for informing council decisions as well as being linked to service improvements. In other words from the local government perspective, participation helps them to function more effectively by providing valuable information. Crucially, 'the goals of empowering citizens or increasing their awareness were largely secondary to the more tangible benefits of improving decision making' (Lowndes et al., 2001a:211). The local authority respondents indicated that ultimately the final decisions on whatever issues were raised via participation should always lie with elected members. Similarly in the US empowerment zones, it was found that city elites took the lead in the programmes that tended to overlook the ability of local communities to restructure themselves (Gittel et al., 2001).

Case Study: UK modernising agenda

Currently, enhanced public participation is central to the UK government's modernisation agenda for local government (Levitas, 1998; Lowndes et al., 2001a). Modernisation is being pursued so that local government can 'in partnership with others, deliver the policies for which this government was elected' (IPPR, 1998:22). New Labour's ethos of co-operation and collectivist approaches rather than individual solutions are realised (Williams, 2003) along with the ideal of community capacity building (Barnes et al., 2003). This is evident through initiatives such as Local Strategic Partnerships that emerged from the Local Government Act 2000 as a statutory duty of local government authorities. The strategies must 'be prepared and implemented by a broad "local strategic partnership" (LSP) through which the local authority can work with other local bodies' (Department of Environment, Transport and the Regions [DETR], 2001:para12). DETR suggests that the key to an effective community strategy will be successful partnership working and community participation. Participation is valued not only for its effects, but also for its inherent value. In this way bringing people together and achieving community interaction and participation can be as valuable as the task that they have come together to do. As the Community Strategy guidance recognises 'the process by which community strategies are produced is as important as the strategy itself' (Office of the Deputy Prime Minister [ODPM], 2000:5). It recognises the value of participation as a means as well as an end. In fact Barnes et al. (2003) claim that enhanced public participation is viewed by New Labour as achieving the following: improving public bodies' decision making, both quality and legitimacy, having the potential to address the democratic deficit and building community capacity and social capital.

The above case study shows how increased levels of participation in the UK are not about citizen control as proposed by Arnstein, but they can be seen as having the potential to contribute to a process of democratic renewal (Lowndes et al., 2001a). Nonetheless Arnstein's model does begin to show how issues of empowerment are not straightforward. In the radical era of the 1960s citizens sought to gain power from authority. However it is not clear how this debate of empowerment and control has progressed within the current era of globalisation, decentralisation and modernisation of local government. The question of which citizens are made powerful remains elusive—communities, women, the poor, socially excluded—whoever these groups may be (Cleaver, 2001). And on what issues and to what extent are they empowered? Taylor's (2003) research discovered that consultants were the only people seen to be empowered as they were able to access money. Indeed the institutional apparatus that is being established to implement the Review of Public Administration in Northern Ireland would hint that many civil servants are being empowered. Meanwhile the experience of and benefits to local communities remain to be determined.

What is clear within this agenda of decentralisation is that the relationship between the state and the governed has altered. The nation-state no longer assumes responsibility for doing things; it cannot afford to deliver services in the way that it did in the past. Citizens in the 21st century demand a different approach to those of previous eras; nation-states typically do not engage in territorial assault or employ coercion tactics. Instead policies are delivered in conjunction and in co-operation with an assortment of agencies and organisations at the local, regional, national and international level. This new governance is depicted by the situation in England. But as we have seen, the experience of decentralisation and the emergent governance structures are not unique to England; the pattern is evident across OECD nations (OECD, 2005). Therefore the analysis that follows, while specific to the English experience, has resonance beyond. It illustrates the complexity of governance, highlighting the multi-scalar nature of these structures.

Rural Regeneration in England

The relationship between central and regional government in England provides an example of multi-scalar governance in practice. Not only has the meaning of rural development changed at the English national and the European level to encompass a more territorial and multi-functional approach (see following section), but there has been a national shift to integrate and also devolve regional policy (Pearce et al., 2005:198). The Single Regeneration Budget (SRB) illustrates the complexity of this model. As part of the devolution process, English Government Offices and the Regional Development Agencies (RDAs) were given an enhanced role in delivering

European and central government policies. Specifically in 1999 following their launch, RDAs inherited a number of funding programmes such as Rural Priority Areas[1] and the SRB, incorporating the Rural Challenge Fund[2]. Taken together these programmes had in turn been governed by a plethora of central government agencies including DETR, Department of Trade and Industry and Ministry of Agriculture Fisheries and Food. Eventually in 2002 responsibilities for regeneration were devolved to the RDAs through a Single Programme, giving the RDAs the autonomy to decide how best to allocate the funds.

Consequently in the early days of their inception, many of the new RDAs were coming to terms with internal, organisational issues[3] while endeavouring to fulfil statutory requirements including administering the previously mentioned programmes and publishing their first Regional Economic Development Strategy. Explicit within this strategy was the requirement to take account of particular rural areas of the region and thereby acknowledge the features that make a particular territory or region distinct. But the RDAs struggled to achieve this differentiation while also integrating the rural issues and interests within a new organisation (Ward et al., 2003). Other matters were impacting on rural policy at this time. Reform of European rural and agricultural policy was also taking place, through Agenda 2000 (see following sections in this chapter). Nationally, the Labour government published the long awaited white paper for rural England 'A Fair Deal for Rural England' in 2000 in parallel with the urban white paper. The document was informed from sources beyond government as it drew on the consultation document *Rural England* published in 1998, with at least 2000 people responding and the government 'listening' to their viewpoints (Department for Environment, Food and Rural Affairs [DEFRA], 2000c: 1.19). Also informing the White Paper was the *Rural Economies* report (Performance and Innovation Unit, 1999).

The summary Rural White Paper states, 'Our guiding principle in both is that people must come first' (DEFRA, 2000d:1). In the full document the government outlines its commitment to rural areas setting out 'what they can expect' (2000c:1.13). In this the government pledges to ensure that rural communities get a fair deal in public services and that the government's rural policies are joined up. Central to and implicit within the paper is the notion of governance in that it sets out the way in which governmental and non-governmental organisations work together. 'Government's role is to provide the framework and support within which people can succeed and the flexibility to develop appropriate local solutions' (2000c:1.13). Onus was therefore placed on the public to help the government achieve its commitment as terms such as *empower, help, work with* and *support* are used to describe the relationship between government and rural communities. It is not simply about the government pledging to do certain activities; success will only happen with the support of rural communities. The government's overall 'goal is to help people in rural areas to manage change, exploit the opportunities it brings, and enable them to create a more sustainable future' (2000c:1.14). The government thus makes it clear from the outset that it will not be doing things for rural communities, but that they have a responsibility to take action.

Assistance is given to communities to enable them to achieve their responsibilities. The 'White Paper sets out a toolkit of measures which local communities can apply to meet their priorities and concerns' (2000c: 1.12). For instance the government vouched to 'strengthen the most local tier of administration, the town or parish council, and give it a bigger role' (2000b:8) and so a parish fund was created. The fund was designed to 'give them more freedom to decide on the kind of help they need' (DEFRA, 2000d: 5) in recognition of the belief that local people are best placed to identify the challenges and opportunities within their community and to respond to these.

The Rural White Paper reflects much of New Labour's policy rhetoric where core ideals such as empowerment, participation, capacity-building and community involvement predominate. Policy rhetoric espouses that responsibility for achieving local solutions does not lie with one particular sector, but each geographical area is allowed to take a flexible approach reflecting local needs and priorities. Not only does this demonstrate the government's commitment to decentralisation, but it highlights the emphasis on a territorial, integrated approach over the established sectoral approach. These principles are embedded in the 'Single Pot' that is administered by RDAs for their particular geographical area.

In April 2005, DEFRA assumed responsibility for all rural policy development; the Countryside Agency's rural regeneration work was transferred to RDAs. Meanwhile resources to support the rural voluntary and community sector are largely administered by the regional Government Offices. The National Environment and Rural Communities Act 2006 created a new integrated agency, Natural England 'to act as a powerful champion for the natural environment' and to create the Commission for Rural Communities. The latter agency has three key functions: rural advocate, expert advisor and independent watchdog with a view to tackling rural social and economic disadvantage. 'It will be a powerful new rural advocate unhampered by delivery functions' (http://www. defra.gov.uk/rural/ruraldelivery/bill/default.htm, last accessed 18.08.08), but as it does not have funds to allocate, it could alternatively be viewed as a sop to the powerful rural lobby. Sparked by the ban on hunting with dogs, the 2002 Countryside Alliance Liberty and Livelihood demonstration through central London grew to become a larger protest against a range of problems that rural communities claimed to face following the foot-and-mouth crisis including prices paid to farmers for produce like milk and school closures (http://news.bbc.co.uk/1/hi/wales/2271393.stm, accessed 18.08.08). The protest left many policymakers and politicians unsettled as they had not anticipated such a strong reaction.

Following the initial phase of devolution in England, decentralisation has progressed. Consequently the mechanisms for implementing rural development and regeneration programmes have evolved to reveal a complex web of governance. The RDAs have responsibility for national rural economic and social regeneration programmes, including this component of new European programmes. However the Rural Development Programme is being led overall by the Department for Environment, Food and Rural Affairs in conjunction with Government Offices in each region. Perhaps most revealing within the new institutional arrangements is the fact that the RDAs are sponsored by the Department for Business Enterprise and Regulatory Reform (rather than the Department for Environment, Food and Rural Affairs which has responsibility for the England Rural Development Programme). Further, rural economic and social regeneration functions remain with the RDAs as clearly stated in the Regional Development Agencies Act 1998, so that 'a regional development agency's purposes apply as much in relation to the rural parts of its area as in relation to the non-rural parts of its area' (Regional Development Agencies Act 1998, 4.2). This is echoed in the implementation of the new English Rural Development Programme, where the RDAs assume responsibility for delivering the economic and social elements, with agri-environmental schemes being implemented by Natural England. Sectoral interests and the centralised administration of them appear to remain strong despite rhetoric of integrated rural development. We will return to this issue in the discussion that follows.

RURAL AND AGRICULTURAL POLICY

As economic growth occurs across the globe, increasingly nation-states are shifting from an agrarian base to reliance on a service sector economy and are residing predominantly in urban areas. This has been happening for some time in Western Europe and we continue to see the shift among eastern European states, the newest members of the EU today. Further east, China and India are coping with the socio-economic ramifications of this phenomenon. Whereas in the past people were often a single generation away from earning a livelihood on the land, today young people are typically far removed from agriculture. It could therefore be argued that their empathy with the farming community is reduced as direct connections with the land are eroded. Of course, exceptions to this exist, as is the case in France where the farming lobby remains the strongest in Europe. Even so, food is increasingly perceived as a consumer good only: more the product of industrialised agriculture and less the product of a farm family business. The most recent global credit crunch brings the topic of food and agriculture to the centre stage, with consumers questioning the perceived high cost of food in the weekly household budget. The Food and Agriculture Organisation (FAO) calculates that in 2007 the food price index rose by nearly 40%. It is estimated that in 2005 the CAP cost UK consumers £3.5 billion through higher prices (http://www.number10.gov.uk/Page15332 last accesssed 15.08.08). Worldwide the effects of the increase in food prices is evident, although not all countries have witnessed the fatal riots of Haiti in 2008 where the prices of rice, beans and fruit rose by 50% in one year (http://news.bbc.co.uk/1/hi/world/americas/7331921.stm, accessed 15.08.08). The critical nutritional situation of poor people and serious inflation have been felt in many countries including Mexico, Yemen, Bangladesh, Morocco and Egypt. While nation-states continue to play a key role in developing policies, increasingly global organisations assume an important role as this excerpt from the G8 Leaders' Statement on Global Food Security reveals:

> We support CAADP's (Comprehensive Africa Agricultural Development Programme) goal of 6.2% annual growth in agricultural productivity, and work toward the goal of doubling production of key food staples in African countries meeting CAADP criteria in five to ten years in a sustainable manner, with particular emphases on fostering smallholder agriculture and inclusive rural growth. (G8 Summit 2008, Tokyo http://www.g8summit.go.jp/eng/doc/doc080709_04_en.html, last accessed 15.08.08)

Internal European political pressure to reform agricultural policies has been strong; this has to some extent been surpassed by external factors, most notably the World Trade Organisation. The influence of globalisation is intensifying. The Uruguay World Trade Agreement marked reduced global protectionism and it is anticipated that the Doha Development Round, which is in its seventh year, will result in nothing more than a minimal change, if indeed it ever reaches agreement having failed to do so in July 2008 (the problem has been pointed to the five-year programme of agricultural subsidies that was recently passed in the US and was described by EU Commissioner Peter Mandelson as 'one of the most reactionary farm bills in the history of the US' [Beattie and Williams, 2008]). However the rising voice of the emergent economies of the G20 nations including India, China, Argentina and Russia will continue to exert pressure on the reduction of barriers to trade in future negotiations. Meanwhile as these emerging economies mature and their population becomes more urbanised and consumer driven, tastes and aspirations are likely to evolve to mirror the lifestyle of citizens of the so-called West. This will increase demand for particular foodstuffs such as wheat, placing ever greater pressure on the global food (and agricultural) market.

The current food crisis underlines the political nature of agricultural and food policy. It also shows how few issues can be considered in isolation. Environmental concerns and rising energy prices intertwine to produce a complex global problem. Extreme drought in Australia, one of the world's largest wheat producers, has had an impact on global wheat production. This affects almost all African countries as they are net importers of cereal (von Braun, 2008). At the emergency food crisis summit in Rome in 2008, the FAO director general asserted that global policies on food security favoured the West. Specifically the issue of subsidies for biofuels is perceived to be one of the reasons why food prices have increased substantially as land is taken out of world food and animal feed markets. There is no agreement on the contribution of biofuels to increased food prices; the US claims that it contributes 2% to 3%, while the International Monetary Fund estimates that 20% to 30% of food price increases during 2006–2008 were due to biofuels (Borger, 2008). And so while statistics paint a particular picture, politics and spin remain key tools for international politicians as they engage in structures of governance.

Political pressure has always had a huge impact on the direction of agricultural policy. Smarting from experiences of food shortages and rationing, emphasis on European agricultural policy in the post-war era of the 1950s was on increased production and self-sufficiency. Things have since changed with the advent of surpluses and the engagement in global trade agreements. In Australia and New Zealand subsidies for agricultural and food products have been reduced since 1970. In the US and Europe there is ongoing tension between politicians and various stakeholder groups regarding the levels of subsidies and tariffs within the Farm Bill and the Common Agricultural Policy respectively. For instance the current 2007 US Farm Bill, also known as the Food, Conservation, and Energy Act of 2008, causes controversy among environmental groups as it subsidises the production of biofuels and accelerates their commercialisation. This has implications beyond with its impact on food production and the ramifications for families with low incomes. Meanwhile in contrast to the heyday of the 1960s and 1970s, today Europe's trade balance is negative. The most recent reform of the EU's CAP, the so-called Agenda 2000 reforms, represents the culmination of a number of pressures on Europe's most expensive policy. With nearly 50% of the budget being spent on supporting farming activities through the CAP, it is inevitable that since its inception in 1962 it has been subject to many and varied reforms. The following section focuses on the evolution of Europe's agricultural and rural policies.

The Common Agricultural Policy (CAP)

The European agricultural lobby has traditionally been strong, seeking a very protectionist policy. This is reflected in the Treaty of Rome (1957) which stressed the importance of a stable and efficient agricultural sector as it aimed to

- increase production by promoting technical progress
- ensure a fair standard of living for the agricultural community
- stabilise markets
- assure availability of supplies
- ensure that supplies reach consumers at reasonable prices (Article 39).

This laid the foundations for the CAP which has evolved, grown and been subject to many different reforms. It adopts a mercantilist approach using mechanisms including price support, import taxes, export subsidies, production control and intervention to ensure that European farmers have protection from global markets. Over the decades the pressures on the CAP have been immense; they have come from diverse sources and these have not always converged. While the Cold War emphasised the need to secure food supplies, the visibility of the food surpluses during the 1980s and the contrasting television pictures of starving children in Africa resulted in political pressure to curb production. For instance as part of an ongoing programme of reforms, milk quotas were introduced in 1984; they look likely to be scrapped completely in the near future.

Meanwhile the publication of Our Common Future in 1987 brought the notion of sustainable development to the world stage. While sustainable development pays attention to economic, social, environmental and cultural matters, it is true that it is most often associated with environmental issues. Five years later the supremacy of environmental interests was evident in the Earth Summit convened in Rio where over 100 Heads of State signed up to the agreements which encompassed issues including biological diversity, climate change, forest management and a blueprint for sustainable development, Agenda 21.

More recent political pressure has heightened demand for a transparent system, with an increase in compensation payments through schemes such as the Single Farm Payment and a reduction in opaque market support mechanisms. Budgetary pressures have also been evident. Taxpayers are not willing to subsidise an economic sector; they are persuaded by the concept of the free market that they are immersed in and of the choice and competition that accompanies this system, and secondly they cannot relate directly to the notion of food shortages or to food security. The concept of the free market and liberalisation permeates global trade negotiations and as a result price support packages are reduced as attempts are made to reconcile them with world food prices.

The inclination to liberalise extends to the expansion of Europe to the east. As this occurs the potential for the financial burden to drastically increase is great given the sheer numbers of people directly involved in agriculture. For instance even following the post-communist reform, when Poland joined the Union in 1994 one quarter of the workforce was engaged in agriculture, while two million private family farms were in existence (Ingham et al., 1998). It was estimated that overall the expansion of the Union at this time by the accession of eastern European States threatened to increase the CAP budget by two thirds (Denny, 2001). But the argument goes beyond a purely economic one with welfare issues to consider. These transition countries face major change as their economies industrialise at a rapid rate, and so there is a counter argument that suggests that support for agriculture during this challenging phase is not only shrewd in the interests of well-being, but necessary to ensure the creation of vigorous economies. The issues associated with the CAP are evidently multi-faceted.

Meanwhile it is argued that public health is jeopardised by the CAP with links being made to the development of major diseases such as obesity and high blood pressure (Birt, 2007). Arguments for the serious reform or scrapping of the CAP are made on other grounds including social justice. For instance the subsidies for biofuels have been accused of acting as an implicit tax on staple foods, on which the poor depend (von Braun, 2008).

Reform of the CAP will always be tricky, but given the high levels of 'support dependency' within European agriculture and the fact that it is completely embedded within the cost structure, Harvey maintains that 'the greater will be the resistance to its removal' (2004:267). And so it remains certain that the CAP will be around for some time to come. But we can be equally assured that it will be subject to ongoing reform as socio-economic and political matters continue to exert pressure. These tensions are evident in the Agenda 2000 reforms which are viewed as setting out the framework for the eventual transition of the CAP to a truly Integrated Rural Policy. However it has been noted that it was an opportunity missed for genuine transformation with a compromise between market liberalisation and protectionism (Lowe and Brouwer, 2000; Lowe et al., 2002). Nonetheless a number of significant changes within the CAP are worthy of note; firstly, as we have seen, the international arena is set to play a growing role and secondly the second pillar of the CAP will assume greater importance (Petit, 2008). Consequently, and of direct relevance to this book, there will be an increased number and diversity of actors involved in the policy governing process. Given the reliance of the CAP on the apparatus of governance, such as partnership and participation it is clear that a better understanding of these issues will inform future rural development policy and practice.

These matters will be scrutinised throughout this book.

INTEGRATED RURAL DEVELOPMENT

In addition to the hundreds of thousands of small groups operating locally and across borders, Keane (2003) describes a tectonic increase in the number and variety of international non-governmental organisations, estimating that 50,000 international non-governmental organisations operate at a global level. For example in Mexico, the micro-regions strategy directs action through 263 Strategic Community Centres based on priorities that have been agreed through a participatory and integrated approach within the local community (OECD, 2006). Social fund projects in Armenia, Peru and Zambia are typically managed by project-implementing agencies which are locally elected with a mandate to act on behalf of the beneficiary community (Babajanian, 2005). Policy coherence around rural issues between ministries in Canada is secured through its 'rural lens' approach. Meanwhile its Community Futures Program advances bottom-up economic development in rural areas. This is similar to the objective of Europe's LEADER programme which represents an integrated, endogenous approach to rural development. Consequently an array of newcomers has appeared on the new rural governance scene, including active communities and active citizens (Murdoch and Abram, 1998), along with agencies and organisations from beyond traditional formal government structures, such as Housing Associations (Goodwin, 1998).

These programmes epitomise the OECDs New Rural Paradigm which is characterised by six features:

- A shift from an approach based on subsidising declining sectors to one based on strategic investments to develop the area's most productive activities
- A focus on local specificities as a means of generating new competitive advantages, such as amenities (environmental or cultural) or local products (traditional or labelled)
- More attention to quasi public goods or "framework conditions" which support enterprise indirectly
- A shift from a sectoral to a territorial policy approach, including attempts to integrate the various sectoral policies at regional and local levels and to improve co-ordination of sectoral policies at the central government level
- Decentralisation of policy administration and, within limits, policy design to those levels
- Increased use of partnerships between public, private and voluntary sectors in the development and implementation of local and regional policies (OECD, 2006)

In order to understand the intricacies of an integrated rural development approach, European rural policy will be examined in detail. As will become clear, the origins lie outside of agriculture per se, but as agricultural reforms have been implemented and rural policy has progressed, their paths have crossed and consequently Europe is now in a position where the CAP concentrates on both.

European Rural Policy

Integrated rural development has its origins in community development practices of the 1950s and 1960s. These were typically used in the context of aid to the so-called developing nations of Africa, Asia and Latin America. The central ethos of these programmes was to stimulate the creation of initiatives from within the community with a view to improving the quality of life of the whole community (Morris, 1981). Although this approach was discarded primarily due to its limitations such as its inability to become self-sufficient, over-reliance on local notables and external experts and inadequate levels of participation, the EU rural development programmes of the late 1980s bear the hallmarks of these colonial schemes. This is encapsulated in the LEADER Initiative which has recently been mainstreamed into the Agenda 2000 reform of the CAP. Before this scheme is examined, some background to Integrated Rural Development is provided.

Ten pilot projects that comprised the Integrated Development Programme were launched in 1982 (Shucksmith, 2008) and marked a shift towards a territorial, integrated approach to rural development. Building on this and mindful of the inadequacies of the CAP and of the plight of rural areas, the Commission published 'The Future of Rural Society' (CEC, 1988). In this document it outlined a strategy for the development of lagging regions to achieve social and economic cohesion. But this was to be done from within the locale so 'that the involvement of local and regional authorities and other social, local and regional economic interest groups in the identification of problems and the quest for solutions limits the number of errors of diagnosis that are all too common when planning is carried out from the outside' (CEC, 1988:62). So not only was this a governance approach in terms of the different and multiple levels of organisations involved, but it embodied an integrated and territorial approach to development recognising the role of the region in a sustainable rural development policy. Furthermore it emphasised the role of local people in identifying solutions for their area. This was unlike traditional agricultural policy which was sectoral by nature. Structural funds were used as the mechanism to allow areas to negotiate with the Commission on suitable strategies for their area. These were allocated according to need measured by deprivation indicators (Ward and McNicholas, 1998).

Later in 1995 Franz Fischler, the European Agricultural Commissioner, assembled a gathering of European politicians in Cork that led to the publication of the Cork Declaration. Fischler reopened debate on agricultural reform. The Declaration acknowledged the diverse needs of rural areas and the requirement of different policy responses. Although it was ambitious, seeking to reduce and simplify the plethora of funds and schemes and advocating a 'much-expanded rural development programme to embrace the whole farmed countryside', it was not ultimately endorsed (Lowe et al., 2002:2). The Cork Declaration was a prelude to the reforms that culminated in Agenda 2000 and the associated manoeuvring illustrate the political nature of European agricultural and rural policy.

It does however embrace the notion of multi-functional agriculture which recognises the responsibilities of farmers as stewards of the countryside and the assets contained therein. As a result a range of outputs emerge from farming activities such as food quality and safety; agri-tourism; and environmental amenities. Multi-functional agricultural is not uniformly understood (Wilson, 2007). Broadly it has emerged as a result of liberalisation aspirations, environmental and food security concerns and the negative externalities arising from intensive farming such as food safety and public health. Multi-functional agriculture is about fulfilling a number of functions for the region and for society more generally. By definition it involves the production of joint goods and part of this corresponds with the needs and expectations of society. It is a 'proactive development tool to promote more sustainable economies of scope and synergy' (Marsden, 2003:185). Typically both a public good and a private good are produced as a result of multi-functionality (OECD, 1999). The prominence of multi-functional agriculture was evident at the Rome Food Summit when Brazil's officials were defending their country's cultivation of sugarcane ethanol; part of the way in which they did so was by 'distributing glossy brochures extolling the fuel's environmental and social benefits' (Borger, 2008). Europe's Agenda 2000 reform policies are seen to support multi-functional agriculture as they aim to reduce agricultural support and high levels of food production while diverting funds into environmental actions. But the reality is of some tension between the multi-functional ideology and of the retailer-led food supply system that relies on a state-supported agri-industrial model (Marsden and Sonnino, 2008).

The first pillar of the reformed CAP is concerned with market support and direct payments, namely the Single Payment Scheme. But evidence from both the US and the EU suggests that current subsidies-based policies are ineffective in addressing socio-economic challenges facing rural communities and also have an uneven impact across an area (OECD, 2006). Correspondingly the second pillar refers to expenditure under the Rural Development Regulation and this is aimed at helping rural communities to develop and diversify. In accordance with rhetoric of territorial bottom-up development that the Commission espoused during the late 1980s, each member state is obliged to draw up a seven-year rural development plan at a geographic level that it deems appropriate. This provides a framework for the allocation of funds, sourced from the EU and the member state. At the heart of these plans is the LEADER approach and a reliance on participation, involvement, decentralisation and partnership.

The LEADER Approach

LEADER was established by the European Commission in 1991 and represented a very new policy style with the unique combination of small, local groups, innovative development projects and flexible funding (Ray, 1998). Recognising the limitations of the colonial community development style approach whereby experts parachuted into an area, the LEADER initiative placed emphasis on an integrated, bottom-up approach. The policy rhetoric was one of empowerment, participation, community, capacity building and finally of innovation as the programme was considered a rural laboratory. The Commission used LEADER to devolve some responsibility not only for the design of rural development programmes, but for implementation. Partnerships comprising representatives from the voluntary and community, public and private sectors formed Local Action Groups. They devised area-based strategies from which individual projects could be funded and as such took the lead in running the programme within a particular territory. This allowed the territory to cultivate its own 'development repertoire' which paid attention to the specificity of that area and so took account of all aspects of language such as food, craft, language and dialect, landscape and music (Ray, 1999a:525).

There were three distinct phases to the LEADER Initiative, correlating to the different phases of the European Structural funds so that Leader I operated from 1991 to 1994, Leader II from 1994 to 1999 and Leader + from 2000 to 2006. The programme has been superseded by the Rural Development Regulation within pillar two of the CAP. This is significant as it denotes the importance of the governance approach.

It is within this aspect of the CAP that regions within member states have discretion to set out their plans for expenditure under their rural development plans (2007–2013) drawing from a set of activities set out in Article 33 of the Rural Development Regulation. These are far reaching and include traditional agricultural activities as well as broader rural development measures. For example among the thirteen listed are the development of villages and protection and conservation of the rural heritage; land improvement; diversification of agricultural activities; and establishment of farm-relief services. So for example the NI Rural Development Programme (NIRDP) was approved in July 2007 and has four key themes:

- Improving the competitiveness of agriculture and forestry by supporting restructuring, development and innovation
- Improving the environment and countryside by supporting land management
- Improving the quality of life in rural areas and encouraging diversification of economic activity
- Using a LEADER-type approach (http://www.dardni.gov.uk/index/rural-development/nirdp2007-2013.htm, last accessed 13.08.08)

Bryden urges a note of caution, as he argues that although there is a common 'impression that resources for 'rural development' have been increasing, it is not clear that this is the case when looked at in 'real' terms' (2005:6). From a budgetary perspective the Rural Development Regulation (RDR) is small at approximately 10% of the CAP total, although France and the UK have used the opportunities presented by the modulation option to redirect funds from commodity support to rural development and agri-environment activities (Lowe et al., 2002). Even so, it could be argued that in the UK by capping modulation at 4.5% instead of the EU maximum of 20%, this is a hollow nod in the direction of reform lobbyists. In this way European agricultural policy has become more decentralised as horizontal measures allow member states to shift a limited amount of funds from pillar one to two; that is, from direct farm support to rural and environmental measures.

But set against the vociferous claims of integrated rural development and of the accompanying accoutrements of partnership, community involvement, capacity building and of bottom-up development, some critical voices can be heard. The specific challenges of the LEADER initiative have been extensively documented and include issues of tension between economic and social objectives; legitimacy of partnerships; capacity building and pre-development; weak consultation processes; incoherence between layers of governance; lack of strategic direction and unrealistic time pressures (see Scott, 2002, 2004; Shortall and Shucksmith, 2001 for a full discussion). Further Stevenson and Keating (2006) suggest that to some extent agricultural interests have recaptured European rural policy simply because they were left with little choice due to budgetary restrictions to traditional agricultural activities, historical negotiating structures that reduce the participation of other rural actors and the mainstreaming of LEADER so that it is in danger of becoming an agricultural initiative. Returning to the example of the NIRDP, early indications show that some members of the farming community perceive this money to belong to their sector. At a public consultation meeting to attract new members to the new Local Action Groups, the bodies charged with devising a strategy for an area and supporting local projects accordingly, a farmer made his position clear:

> 'Well I would like you to guarantee me, as a farmer that this money is not all going to be spent on rural development. We [the farmers] face tough times at the moment and we don't want this money being spent on projects, it needs to be invested into proper farming activity' (27.02.08, Farmer attendee to public consultation).

Sectoral policies and the centralised sectoral administration of them remain very important (Bryden, 2005). As a result there is evidence of contradiction between policies at the different scales of governance. In Northern Ireland sectoral policies such as education or health continue to be made by distinct departments. Consequently while the NIRDP sets out its aim to improve 'the access by rural dwellers to basic services for the economy' (NIRDP, 2007:95), a recent report on the Education sector in Northern Ireland recommends that 'the minimum (not optimal) enrolments for new primary schools . . . should be (i) Primary: . . . 105 pupils in rural areas . . . When the enrolment in an existing school falls below the relevant level, the future of the school should be reviewed' (Department of Education for Northern Ireland, 2006:31).

In accepting the recommendations the government minister subsequently makes it clear that 'this is not an agenda to close small schools' (Strategic

Review of Education: Government Response of 12 December 2006). Nonetheless there is evidence of small schools closing across rural Northern Ireland (http://www.northernireland.gov.uk, last accessed 15.08.08).

So while integrated rural development embraces admirable ideals and
seeks to bring about positive change for all rural residents, the reality
appears to be more complicated. Even though a bottom-up approach is
inbuilt to territorial rural development programmes, different sectoral
interests appear to have the capacity to exert pressure on the development process. There is tension between the objectives of top-down and
bottom-up processes. Further centralised administrative regimes continue
to advance particular sectoral interests, despite a broader policy discourse
of devolution and decentralisation. These matters are played out within
structures of rural development, typically a partnership arrangement. This
book will examine the way in which different interests and representations
are manifest in rural development policy and practice.

3 The Case Study

This chapter provides essential context for, and a description of, the research. The particular structures for the regeneration activity are presented. The chapter progresses by analysing methodological issues that are significant to the specific aims of the research. These are matters that should have resonance with rural researchers and also with practitioners and policymakers.

The institutional configurations for rural policy in the case study reflect many of the developments that were apparent within rural development and regeneration. During the 1990s the area was a recipient of European structural funds (Objective 5b) and at this time Leader I and II regions were designated within the area. In addition it encompassed a Rural Development Priority Area as designated by the then Rural Development Commission. The result of these classifications meant that the area had a history of undertaking development initiatives at a community level and also of accessing national and European funds. Various partnerships bringing together voluntary, community, statutory and private sectors existed to address local need and to bring about positive change. A vibrant rural development and regeneration sector existed and included active community councils, local authorities, charitable bodies and voluntary organisations. Many of these organisations were connected through networks and umbrella bodies and were familiar with working together through a partnership approach.

In the section following, the rural development project is described in some detail, along with an account of the communities involved. Firstly the host organisation is presented.

HOUSE

I conducted this research over a three-year period while employed by a housing association (referred to as House) specifically to co-ordinate a rural development project, the Community Initiative, which was sponsored by two UK government agencies.

At the time that the research was conducted there were approximately sixty employees at House and four stand-alone 'good practice' projects. Compared to many other housing associations, it was a fairly small operation. However, the organisation took pride in the fact that while it was small in size; its impact was not unsubstantial. It aimed to initiate new approaches to affordable housing and to establish schemes not directly concerned with housing; these will be reflected on later in the chapter.

As with most housing associations, House is managed on a daily basis by a senior management team (SMT), headed by a managing director and assisted by deputies. In addition to the staff, a committee, which meets monthly, provides organisational management and governance by discussing strategic matters such as housing development schemes, financial audit and rent policies.

Housing associations are managed and monitored by a government agency, the Housing Corporation, and they receive grants to build houses directly from government. Providing housing is their core business. The Housing Corporation, the funding and regulatory body for housing associations in England and Wales, holds a Public Register of Social Landlords (housing organisations). This register illustrates the wide range of housing organisations that currently exist. At one end of the spectrum are agencies such as The Places for People Group which 'is one of the UK's leading housing and regeneration specialists and is responsible for more than 47,000 homes in England, Scotland and Wales' (http://www.placesforpeople.co.uk/index.aspx, last accessed 07.07.08). Meanwhile at the other end of the housing and regeneration continuum are associations such as Highbridge Society Limited which manages ten homes in Sussex (http://www.housingcorp.gov.uk/server/show/ConRSL.579, last accessed 07.07.08). Within the context of housing provision and regeneration in England and Wales, House is a small provider. At the time of conducting the research it managed in excess of 2500 homes.

Most housing associations specialise in a particular area within the broad spectrum of housing, such as providing homes for low-income individuals who live locally, as was the case for House. Providing housing is a complex task. It involves identifying sites and then building houses on them. This is the job of the development team, it works with a range of bodies including planning agencies, community organisations, parish councils, private landowners and building contractors to identify sites, secure ownership and planning permission and finally to build the houses. The other aspect of housing provision consists of maintaining and managing the houses when they are occupied by tenants. Associations either directly employ maintenance staff or subcontract the whole maintenance process to relevant agencies. Housing management also concerns the tenants' ability to pay rent. A housing manager may work closely with the tenants to help them manage their personal finances so that they are able to meet the obligations set out in their tenancy agreement. Often this involves working in conjunction with other agencies, such as citizens' advice bureaus. Inevitably, at certain times housing managers get involved in tricky family situations, such as with issues of domestic violence or marital conflict. From a short-term managerial perspective, it is the job of the manager to ensure that the rents are collected for the association. However, housing managers tend to work with the tenants in whatever way they can to ensure their ongoing tenancy. This may mean providing links to marriage guidance agencies or helping them to set out a budget plan.

Beyond Housing Provision

During the late 1990s, housing associations were placed under pressure by central government to diversify their remit, and so many embarked on regeneration activities. The pursuit of these other functions meant that housing associations had to attract funding from other sources such as regeneration programmes. In fact the Housing Corporation also had a special initiative, the Innovation and Good Practice scheme, that had been established to encourage the development of novel approaches and activities by housing associations. Consequently House was involved with a number of activities including pioneering the use of private funding in building affordable homes, promoting shared ownership, introducing energy efficiency in affordable housing and pertinent to the subject of this book, the Community Initiative. As part of this process, it maintained a very strong network of contacts among policymakers and had impressive links with budget holders in relevant organisations. The organisation was extremely proud of its reputation as an innovator. For instance, in one report the chairman identifies innovation as one of three themes for that year. A subsequent annual report claimed that

> 'the gap between the need for affordable housing, on the one hand, and the provision of homes on the other, grows wider and wider. These are challenging times. With our proven track-record for innovation, [House] is perhaps uniquely placed to seek and find new ways to bridge that gap. We have set ourselves ambitious targets and will display the courage to deploy our resources financial and human, flexibly and effectively. Putting it in a nutshell, [House] intends to 'Aim high, keep its feet firmly on the ground and walk tall'. The times are not just challenging. They're exciting' (House Annual Report).

Despite this effusive rhetoric, its reputation for innovation was not endorsed by everyone. House was viewed by some professionals as paternalistic and not necessarily innovative, just 'well connected' (Council officer, 25.02.02). The *[House] family* was a term often used by the SMT. This perception of the working relationships reflected earlier days in the organisation's history when it was smaller and all members of the SMT took a direct interest in all its activities. That culture still lingered in the organisation and was often in tension with the evolving structure that was much less familiar and more managerial. Notwithstanding the emerging managerial ethos, members of the SMT took a keen interest in the Community Initiative. Some of the regeneration and development agencies were surprised that House was leading the project and 'there was suspicion of the project more widely in the community' (project evaluation, 22.01.02). Nonetheless, evidence of House's success in managing projects was found in its extensive publications list, based on previously conducted research and also on specially commissioned work. This involved physical and community-based aspects of house building. Meanwhile House's ability to make connections was unmistakable, as the following excerpt from a note circulated by a member of the SMT to various staff members illustrates:

> 'I spoke to Y after his speech to say that one large element was missing. How do professionals in our position change the culture of our organisations so that we listen to people and enable them to become part of the solution?' (17.05.01)

At this time Y was a political advisor to the prime minister and worked in central government's Social Exclusion Unit.

THE RURAL DEVELOPMENT PROJECT

In early 1988 House compiled a position paper on market towns in England. This was the culmination of a series of meetings held with partners over the preceding two years to consider how best to develop the project. The partners included rural development organisations and funding bodies that then went on to became the steering group for the Community Initiative. The fundamental objective was to demonstrate how housing associations can contribute to rural sustainable development and regeneration.

The steering group included representatives from the main project funders and so it was no accident that the project explicitly met with their objectives—it had in part been shaped by them. It met on a quarterly basis and literally provided a steer for the project. Often members would offer useful links and contacts for the project co-ordinator, and frequently the meetings would entail a discussion on topical issues within rural development. As a group they did not get heavily involved in the daily project activities.

The research was conducted in five different communities during the time period 1999–2002 and included the following:

- The Village: a deeply rural village cluster
- Great Villham: a London overspill settlement

- Growthville: a growing village
- Commuterville: a commuter village
- Market Town: a market town

Thus each area had different features and characteristics and so the project activity varied across the areas. Outputs included successful Single Regeneration Budget bids, community appraisals, community play areas and improved health facilities.

The Village

Located in the centre of the county, The Village has the appearance of a traditional county village, the main street featuring pink cottages alongside Tudor buildings. There are no large market towns nearby, and so many of the surrounding villages use The Village for facilities and services, only going further afield when necessary. Public transport is inadequate, and most people rely on private transport for their journeys. No major roads pass through or are close to The Village; it is surrounded by a network of narrow, often single tracked, roads.

Consequently, although The Village has a population of just over 1000 people it has many facilities. These include a high school, primary school, bowls club, medical centre, tennis courts, football field, cricket pitch, children's play area, fire station, indoor swimming pool, gym and community centre. In addition there are several shops, including a post office, bakery, butcher and hardware store, a couple of churches and three pubs in the village. There is a mixture of housing provision made up of council estates, old and new middle-income private estates and a few luxury housing units. Although some residents work in agriculture and agricultural related sectors, many travel to large towns in the region for service-based employment while some commute to the city of London.

Great Villham

Great Town and Great Villham are adjacent settlements found south of the county capital, with reasonable proximity to London. Approximately 8000 people live in the village of Great Villham, which is made up of an older Victorian style village and a newer overspill settlement. The older part of the village consists of predominantly Victorian housing with facilities including a church, pub, barbershop and newsagents. These are inadequate for the population, and most residents travel to neighbouring Great Town for their weekly provisions. Many community activities revolve round the local community centre, also built during the 1960s. In addition to a primary and upper school, the village has a theatre located at the upper school.

Overspill housing for London was built during the 1960s causing the population to expand greatly. The resulting accommodation consists mainly of short, parallel terraces of four to six fairly small houses. At the time the research was conducted the local council was responsible for these houses and was on the verge of completing an upgrade on many. The remainder of tenants purchased their homes under the government's 'Right to Buy' scheme.

Adjacent to Great Villham is Great Town with a population of just under 12,000 people. It is a thriving market town with a variety of shops in the centre and many other facilities round about, including a leisure centre, theatre, restaurants, pubs and churches. Out of necessity most residents in Great Villham go to Great Town to access services and facilities. In addition the town acts as an important employment, shopping and cultural centre for many more people from the rural hinterland.

As a result of its geographic location many residents commute to London and so their salaries, which are high relative to local earnings, mean that house prices are pushed beyond the reach of many residents. Typically the more affluent choose to live in Great Town and as the town attracts more residents, house prices are raised further, more people are drawn to the town, and so the cycle continues. Great Villham has a scruffier appearance to its pristine neighbour, Great Town which has streets and buildings evoking an idyllic 'chocolate box' image.

To the stranger passing by it would be difficult to identify the dividing line between Great Villham and the town of Great Town. However, there is no mistaking the existence of a difference, least of all in the private housing market with houses in Great Town commanding higher prices than their equivalent in Great Villham. The local economy was considered to be 'extremely fragile and over-dependent upon a large branch plant automotive engineering facility' (Single Regeneration bid, p. 4). The relative prosperity enjoyed in the surrounding hinterland and elsewhere in the area had a masking effect. Consequently a general lack of long-term investment resulted in a poor transportation network, lack of modern healthcare facilities and inadequate sewerage and drainage systems.

Growthville

Growthville lies just off the main road system in the county, a region earmarked by the Regional Development Agency for future development. A dispersed population of 3000 live in a variety of accommodation—a council estate, affordable and luxury housing, as well as mid-range homes, many of which were built during 1970s. A regular bus service provides connections to the nearby town. Unlike many other villages in England, Growthville faces rapid growth and so the residents endeavoured to secure more facilities and services. A new library was built and completed in 2001.

The village contains a post office, garage and car showroom, bakery, several butchers, as well as a railway station, primary and upper schools, a community centre and a sports pavilion. These are spread around the village, making it difficult to locate a central focus.

Commuterville

Containing many colourful and rendered houses, Commuterville straddles the southern county boundary. The village appears well kept and is fairly typical of the middle- to high-income population residing in this part of the county. It has a total population of about 1300 and has many amenities including a village hall, a couple of churches, several pubs, post office, hair stylists, police station, garage, doctors' surgery and a number of shops. A railway station provides connections to the mainline and so many people can and do commute to London.

A full range of accommodation is found in the village including affordable housing, small luxury developments and mid-range bungalows and houses. Residents seem content with this mix of housing stock as no protests were raised at the high cost of the newest luxury homes built in the village. Balanced against this, House was negotiating with a local landowner and the local authority about the potential to build some affordable homes in commuterville.

Market Town

Market Town lies on the north county boundary. It was important historically as the crossroads of the routes between pivotal ancient settlements. This position, once the source of prosperity, is today seen as the reason for decline as it is no longer a busy travelling route. Market Town has become distant from everywhere.

As with most market towns it has many shops and facilities, including a new supermarket, as well as pubs and churches, a community hall and railway station to support its population of just under 9000 people. A US Air Force base nearby contributes to Market Town's economy.

Community Regeneration Activity

Across the areas various types of rural development activity was evident. In many cases a number of different groups existed, such as one focused on a community appraisal and another on community regeneration. As the account of rural development progresses throughout the book, it will become clear that these approaches tend to be led from within the community, or implemented from outside by an agency active in the area. Consequently a range of different activities occurred in each of the areas including the following:

- Creation of housing task groups and completion of housing needs surveys

- Additional healthcare facilities
- Successful SRB (Single Regeneration Budget) projects
- Village Design Statements
- Successful applications to the Countryside Agency's Vital Villages Scheme
- Creation of a community drop-in centre
- Young persons' skateboarding project
- Establishment of new group to manage the refurbishment of the play area
- Community Action Plans conducted

In all cases the local government body, that is, the town or parish council, either led or endorsed the community activity. As a result the membership of local groups included a range of individuals and representatives including parish councillors, representatives from local groups (Women's Institute [WI], environmental group, the school) and external agencies such as health authorities, the local authority, housing associations and the county-wide rural development agency.

Creating Structures for Regeneration

Two main types of rural regeneration were evident, and while all were pursuing strategies that relied throughout on a degree of local participation, they have resonance with the bottom-up and top-down approaches to development (see Chapter 1 for further discussion). The SRB projects in Great Villham and in Market Town are examples of a local government-led initiative whereby programme objectives provided an over-arching framework. Other rural development activity was initiated from within the community and as such a locally based agenda provided the stimulus. In each of these areas one of the key activities that the local group sought to undertake was a community appraisal. While we will not be scrutinising all community action in all of the areas, it is worth noting that various groups existed in each district and so residents had a range of options for participating in community-based activity. An overview of the SRB and of the Community Action Plan programmes is provided in the following section to highlight the context within which the different projects were working.

The Single Regeneration Budget (SRB)

The Single Regeneration initiative was operational in England from 1994 to 1999. It brought together several programmes from different government departments and attempted to make regeneration funding more efficient. It provided support to local partnerships that sought to enhance the quality of life of local people through a range of activities including the provision of education and skills training; housing; capacity building; and support for small businesses and economic development. The SRB scheme emphasised the central role of local communities in regenerating their area with the aim 'to encourage local communities to develop local regeneration initiatives to improve the quality of life in their area' (DETR, 1997:3). Partnership, competition and hands-off management were seen to be embedded in the SRB scheme. 'Central to the initiative has been an emphasis on a partnership led approach to regeneration whereby interested parties come together at the local level to devise a regeneration scheme and seek financial support through an annual bidding round' (Rhodes et al., 2002:11). The partnerships were expected to include a range of organisations representing the local community, voluntary organisations, the private and public sectors (Department for Communities and Local Government http://www.communities.gov.uk/citiesandregions/regeneration/singleregenerationbudget/221229/, last accessed 07.07.08).

Originally the funding was directed at urban initiatives, with alternatives such as Rural Challenge available to rural areas. Eventually, while the bulk of the fund was channelled towards urban regeneration, it was available to both rural and urban areas. For example 20% of the funding in Round 6 was targeted towards 'other' areas including rural (DETR, 1999b:6).

The then Department of Environment, Transport and the Regions[1] (DETR) issued national guidance focusing in particular on delivery plans, project appraisal and approval, financial guidance, and monitoring and periodic review, but it 'is not intended to be a desk instruction for Partnerships and RDAs (Regional Development Agencies)' (DETR, 1999a:1). A territorial and decentralised approach is emphasised as DETR advised that detailed procedures must be developed between the partnership and RDA, whilst adhering to DETR and thus government legislation. So although the day-to-day management of the scheme rested in the hands of the decentralised RDAs, the SRB programme operated within a framework administered by central government via the SRB objectives, the annual bidding process and the approved economic development strategies.

Community Appraisals

With no given budget at their disposal and limited resources, regeneration groups have little choice about the type of participatory techniques that they adopt. Particular techniques are advocated by support agencies, and they tend to predominate among community-based regeneration groups. For instance the Countryside Agency's Vital Village Scheme offered financial support to communities undertaking a Village Appraisal (this scheme closed in 2004; http://www.countryside.gov.uk/VitalVillages/Index.asp, last accessed 28.08.04). Community Action Plans and Village Appraisals were promoted locally by officers from a rural development agency that worked across the county. There was a network of CAP initiatives throughout rural England.

The Community Action Plan Officers were obliged by their funders to produce a certain number of action plans using a specific methodology. Their achievements were measured not on being able to make positive changes within the communities in which they were working, but by the *number* of community action plan projects completed. So although groups were able to tailor make their questionnaire and spend time deliberating on the process, the success of the overall scheme was measured by the number of plans completed.

Specifically the Appraisal consists of a survey that is carried out by a community group and distributed to every household in the area. Viewpoints of residents are sought on issues affecting their lives. The appraisal process often involved the use of the 'Village Appraisal for Windows' software such as that developed by the University of Gloucester; this approach includes report writing, printing and photocopying costs in the region of £1500 (Spedding, 2003). The officers worked closely with local community groups to help them conduct the appraisal, action plan or similar type of community needs analysis. They 'encouraged' community groups to adopt this approach, in a prescriptive way, suggesting that if a community wished to do anything from refurbishing their play area to building a village hall, it was better done in the context of a Community Action Plan scheme (Community Action Plan Officer, 21.10.99). Indeed the Countryside Agency's application form for funding for community projects asked if the council had carried out an appraisal or survey. There is a danger that the agenda of the community is marginalised. Communities are also more likely to restructure themselves to demonstrate that they fit with government requirements (Atkinson, 2003). This may entail re-inventing the wheel, so that previous models of success are replicated. As a result risky projects are avoided and innovation is stifled.

At the same time as SRB bids were being developed in Great Villham and in Market Town a number of other projects were operational in the areas. Of note were community action plan projects and local government strategic pathfinder partnerships (the latter emerged in the context of modernising local government). These partnerships operated alongside one another, each with different agendas and objectives but relying on the participation of residents of the same area.

With regard to one of these schemes, the district council took a political decision to invest in the Great Villham and Great Town area. At the time of conducting the research, although the village and the town contained 3% of the county's population, 20% of the poorest and most deprived wards in the county were found in this area. Furthermore it was affected by a general long-term lack of investment, lack of modern healthcare facilities and inadequate sewerage and drainage systems. A fatal fire on a council estate marked a watershed. Three different residents' groups placed pressure on the council to invest in their community. The District Council responded positively and two major changes resulted. One was the allocation of £35,000 per council house for refurbishment and fire-proofing, totalling £5m over five years (Council Officer, 25.02.02). The other was the devotion of Council staff to the development of an SRB project to address deprivation within the community and to 'kickstart a larger programme of regeneration' (Council Officer, 25.11.99).

Professional and Voluntary Practitioners

The impetus for the Great Villham regeneration came from outside of the immediate community locale, albeit from the local government body. It was driven by professional interests and political pressure. It contrasts with other activities within the Community Initiative including the Village Project, the Growthville regeneration group and the Commuterville Footbridge group where a bottom-up approach prevailed in that the inspiration for developing projects came from individuals within each of the communities. Many of these projects accessed support from professional rural development staff employed by a range of agencies including health organisations, umbrella community agencies and housing bodies.

The term *professional* is used here to signify an individual who is engaged in an activity as their profession, that is they earn an income from it, and frequently have received some level of training to enable them to practice. This is in contrast to the individual who practices regeneration in a voluntary capacity, but is not in receipt of an income from those activities. Ray (1999b) reminds us that professionals pay attention to the interests of the local community, of funding programme officials and bureaucrats and of local and regional professionals operating in the rural development arena. It is important to realise that these actors do not blandly implement programmes, but at some level they influence the very programmes in the area. As reflexive practitioners these individuals mediate 'between officialdom and the local/personal level' (Ray 1999b:23). One of the risks emerging is that 'behind formal legalisms and political rhetoric, the sociological realities are those of subjects, clients and consumers not of citizens of equal worth and decision-making capacity' (Stewart, 1995:74). And so professionals face work pressures of targets and outputs:

> '[Kate] is being put under a lot of pressure to develop lots of projects . . . this is causing problems for the [Community Project]. The [health organisation's] interpretation of community regeneration is very different to [that of the rural development agency]. I'm not sure how much of partnership working is really about building capacity. It seems to be more about different organisations ticking boxes' (Research journal, 18.12.01).

In addition so-called community champions work within local communities striving to stimulate change; they operate on a voluntary basis, have a wide network of contacts, are pivotal to the development process and consequently are often involved in the range of partnerships that exist in the locality. Community champions are often driven by altruistic reasons or are motivated by the standing that they can enjoy within their area as a result of the successful implementation of rural development projects. The role of individuals in the rural development process will be evaluated throughout this book.

Rural development agencies typically have a general remit to support local community groups through specific activities. Schemes may be small scale, designed as pilot projects or good practice ventures as was the case with the Community Initiative. Other initiatives were available across a geographical region, such as the Community Action Plan project that was led by a county-wide rural development agency. As a result it was more widely available to a greater number of communities. These different initiatives will be used throughout the book to illustrate the practice of rural development.

RURAL DEVELOPMENT IN ACTION

Several issues are significant to this research. Given its focus on the rural development community of autonomous actors and intricate power relations among those actors (see Chapter 4), the objective of revealing micropolitics (see Chapters 5 and 6) within a complex governance framework relying on participatory development (see Chapters 7 and 8), a complex research strategy was required. An approach that facilitated a meticulous analysis of the rural development process and all that that entails while supporting a thorough examination of the wider structures was crucial in order to address the overarching research challenge: to examine the disparities between the policy and practice of rural development.

The research is founded on a particular understanding of the individual. People live in material and bounded structures and locations; these contexts shape their interpretative processes and the meanings that they assign to events (Goffman, 1959; Brewer, 2000). Thus people are meaning endowing; they have the capacity to interpret and construct their social world and setting rather than responding in a simplistic and automatic way to any particular stimuli (Mead, 1934; Cooley, 1942; Blumer, 1969). They are also discursive and can articulate their meanings. That is, they can tell others what they mean by something—a comment, an idea or behaviour—and can suggest the motives behind it. The assumption therefore is that as a result of living in a social context, rural development agents not only know and understand what they wish to achieve, but they can convey this information to others. This is significant if we are to identify and distinguish between the preferences, wants and real interests of actors and to determine the exertion of power and other micro-political processes (see Chapters 5 and 6 for a fuller discussion). Nonetheless the powerless fail to recognise that their real interests are at risk and so make no attempt to defend those interests.

Identifying real interests is not a patronising exercise of one agent judging the needs of another, nor does it simply relate to personal tastes or preferences. Real interests are about an 'intersubjective web of meanings through which a community understands its own problems' (Ron, 2008:3). The real interest of the community is determined through an iterative process by which priorities and issues are negotiated and renegotiated in a dynamic and interactive manner by a group of individuals all with their particular values, interests and preferences.

Therefore and as Porter (1995) argues, understanding the actor's viewpoint, although necessary, is not a sufficient condition for social knowledge, we need to be able to shed light on the relationship between social action and social structure. This follows from Weber's contention that the social researcher must understand the social context within which individuals act. He recognised that a mix of motivations cause individuals to act in particular ways in the social world. Individuals never act in a vacuum but are always considering the impact of their action in the context of a social setting, that is, with others in mind.

> Sociology . . . is a science concerning itself with the interpretive understanding of social action and thereby with a causal explanation of its course and consequences. We shall speak of 'action' insofar as the acting individual attaches a subjective meaning to his behavior—be it overt or covert, omission or acquiescence. Action is 'social' insofar as its subjective meaning takes account of the behavior of others and is thereby oriented in its course. (Weber, 1968:4)

Weber's work showed how such understanding or *verstehen* can be used to research the social world and thereby validate meanings and social action. By so doing he introduced these terms into sociological discourse laying the foundation for naturalism[2] (Elwell, 1996b). Naturalism forms the central tenet of the qualitative methodology adopted within this research as it seeks to understand through quality and meaning. The research was conducted within an ethnographic framework that places emphasis on Weber's concept of *verstehen,* or meaning (Elwell, 1996b). It is true that social life is partially interdependent on the concrete situations and structures in which it exists (Hammersley, 1990; Elwell, 1996a; Brewer, 2000; Denzin and Lincoln, 2000; Gillespie and Sinclair, 2000). Social life and social action involve the mediation between structure and agency where 'agent and structures are not two independent given sets of phenomena, a dualism, but represent a duality' (Giddens, 1984:25). So it is insufficient to consider an individual's actions alone, but the activity of others and of social structures must be considered. This research aims to understand the way individuals make decisions, the way in which meaning is ascribed to particular events and the role of the agents and institutions in this process.

Rural development policy and practice are intertwined as structure and agency. Structures are 'the rules and resources upon which social agents draw when acting' (Porter, 1998:161). However norms and values (i.e. rules and resources) are embedded within the agent rather than in the structure (Giddens, 1984). They exist virtually and so are only made real through meaningful actions; indeed they are produced and reproduced by action. Furthermore rules and resources may be material and non-material (McGrath, 2001). Actors or practitioners operate within a framework that is affected by the actions that they undertake. In turn the actors are constrained and enabled by that framework; they derive meaning from rural development structures. Structure and action are therefore interlinked, forming a circular relationship where 'structure is both the medium and outcome of the reproduction of [social] practices (Giddens, 1979:15). Social power plays a role in the deployment of resources and so Giddens (1984) argues that through actions, all agents have the capacity to draw on resources. Hence structures may be constraining to activity, discouraging particular action; but they also have an enabling role as they allow agents to do things in certain ways.

Ascribing this understanding to structures and agents will enable analysis of the issues at the heart of this study. For instance it will allow close scrutiny of how power is used by actors to use resources in particular ways. This can only be done through an ethnographic framework.

Reflexive Ethnography

Although ethnographic practice is currently diverse (Adler and Adler, 1999; Wacquant, 2003), often instilling rancour (Snow et al., 2003), it is alive and well today. The importance of this approach is evident in rural research (see for instance Newby, 1977; Strathern, 1982; Mayerfield Bell, 1994; Pini, 2004). It is a theoretical and analytical process that is about more than navel gazing or biographic analysis; it is concerned with critically engaging with those being researched to understand what causes and effects influence their viewpoints (Cook and Crang, 1995). Therefore the findings of this research are not just applicable within the geographic area studied; they have relevance to the broader rural development and regeneration sector.

The research brings a novel perspective to ethnography: I was employed to undertake and participate in a rural development project and at the same time I was conducting research on the overall process (see McAreavey, 2008 for a full analysis). Reflexivity lies at the core of the methodology and it provides a way of achieving the insight necessary to analyse rural development from this unique perspective. Reflexivity has been described as the capacity to think back on one's thoughts and activities (Mead, 1934) and is often interpreted as a process that explores inter-subjectivities, that is the relationship between the researcher, the field and the researched (Burgess, 1984; England, 1994).

The methodology follows Brewer's model that allows ethnography to look beyond the immediate setting and make connections to wider societal issues (Brewer, 2000). It follows from Mills' (1959:5) call for a 'sociological imagination' and is based on the 'ethnographic imagination' (2000:51). The ethnographic imagination describes the 'imaginative leap necessary to recognise the authority of ethnographic data' (Brewer, 2000:51)[3] and so responds directly to limitations cited by postmodern criticism of ethnographic research—representation and legitimacy. Brewer's (2000) model establishes the authority of the data and provides a broad context for the analysis of microscopic events. It recognises that people live in material and bounded structures and locations; these contexts shape their interpretative processes and the meanings that people assign to events (Brewer, 2000, Goffman, 1959). It is appropriate for this study as it involves critically engaging with those being researched to understand what wider causes and effects influence their viewpoints (Cook and Crang, 1995; Hammersley and Atkinson, 1995).

In addition to reflexivity, the role of what Aristotle termed *phronesis*, that is, wisdom or practical reason, was vital for conducting the research. As a tool within research it is not a wholly conjectural concept, it is not a mere subjective judgment: instead it attempts to achieve 'excellences' pivotal to the relevant community (Dunne, 1993:10). It places emphasis on the importance of culture, value and power in society (Wagenaar and Cook, 2003). The delicate nature of researching power relations meant that as a researcher I had to understand how to behave in particular circumstances in order to navigate a course of action that interacted between the abstract and the concrete.

Participant Observation

Unstructured, flexible and open-ended participant observation methods were used to focus attention on what human beings feel, perceive, think and do in the field. This was mindful of the process of rural development and thereby allowed a full scrutiny of associated processes. Such closeness to the practice of rural development was imperative in addressing the role and significance of micro-politics, in considering complex power relations and in identifying the wider links to the rural development framework. Researching at this level brought with it proximity to structures and agents. Where appropriate, the identity of these individuals has been obscured through the use of pseudonyms and codes (Hoonaard, 2003).

The use of a field diary and notes and interviews and the analysis of policy and funding documentation facilitated conducting research that not only focused attention on the actions of individuals, but also analysed the social and institutional context within which they existed. It focused on the 'behaviouristic' (Fielding, 1993:162) and was maintained to record events at the lowest level of interference. This was complemented by an array of documents including field notes, meeting papers, funding and policy papers from the statutory and voluntary sectors, newspapers and interview notes. The latter were drawn from twenty-five semi-structured interviews conducted with six community residents; five volunteers active in the voluntary sector; eight public sector professionals[4]; and six professionals working in the community and voluntary sector.

4   Power

Power has been applied to studies of society for many decades. It is used to study problems of everyday life. Extensive sociological and social science literature exists on the concept of power, seeking to overcome the imprecision of the everyday language and leading Latour (1986) to deride it as a pliable and empty term. It is true that adopting particular conceptual frameworks in the study of power represents a political commitment. For instance moral and evaluative objectives can be fulfilled by focusing on individualist concepts (Barnes, 1988). In this way responsibility for consequences, or the lack of them, can be pinned to people within society (Lukes, 2005). This is evident in the way in which Mills (1956) assigned responsibility to individuals for particular events in his analysis of elite powerholders. It is true that different ways of defining power are natural to different perspectives and purposes.

This chapter outlines the theoretical understanding that is assigned to power throughout the book. It is unavoidably detailed in order to reveal the nuances and intricacies of power that will enable a scrupulous analysis of rural development processes. The central role of power to rural development is highlighted before going on to conceptualise the concept. The chapter reveals the personal nature of power relations while also highlighting the significance of structures for these actors as they exert power. Consequently Lukes' theory is used to understand its application in rural development. The analysis therefore considers the three faces of power, real interests and 'free' action of 'autonomous' individuals.

POWER AND RURAL DEVELOPMENT

Rural development is fundamentally about bringing positive change to groups of people within rural communities (Buller and Wright, 1990). By its very definition it can become a conflictual process, bringing betterment for some (Shortall, 1994). A ripple effect ensues, where the consequence of action is felt beyond those immediately driving the process. Typically a particular community, be it geographic or thematic, benefits from rural development schemes, but not all individuals within an area are likely to benefit equally.

Ideally policymakers expect that overall rural development will positively affect individuals and associated communities[1]; otherwise they would be unlikely to justify financial support. But at the heart of the activity are a number of key people who work together as a group. Often referred to as animators, the input of these individuals is pivotal to the success of rural development practice (McAreavey, 2003). They may work as professionals or in a voluntary capacity and they are capable of using personal positions of power to mobilise resources to make things happen within a particular community. The impact of their actions has implications beyond these actors.

Conveniently the community power debate stems from the idea of who holds power and who does not. Central to this tradition is the ideal of a community of autonomous persons whose real or implied consent is crucial. The nature of the autonomous individual must be understood. This is not an unqualified and absolute notion of autonomy as individuals are able to 'have experiences, reason, adopt beliefs, and act outside all contexts' (Bevir and Rhodes, 2006:71). The relative autonomy of individuals means that they are affected by values, traditions, social norms and practices; in other words they operate within a particular context. They have agency so that they are able to act in new ways and to change their inheritance. Fundamentally therefore a rural development agent is free to decide whether or not to support certain activities. To earn legitimacy, the powerholder must be deemed acceptable to those consenting. So the actor seeking to regenerate her community through social enterprise must be accepted by others before their approval or co-operation is offered. This may be in the form of participation in working groups or by providing access to essential resources.

Issues of legitimacy and capacity are at the heart of community power debates. The absence of real or implied consent results in illegitimate power which eventually becomes an obstacle to the achievement of individual autonomy (and so the ideal of community is not realised). If we imagine the case of a community champion, it would be virtually impossible to affect positive change were it not for the underlying perceived legitimacy of that individual. This might stem from her position, motivation and credentials to drive forward the regeneration initiative. This is a fundamental point for our understanding of power. The powerholder's position is not necessarily apparent through the exercise of power. A position of power can be much more subtle. It is fundamentally about having the *capacity and legitimacy* to exercise power, and the perception of this position by others, enabling the powerholder to call on the obedience of others (Lukes, 2005). We return to the issue of power and rural development later in the book, but first the remainder of this chapter critically analyses theories of power.

CONCEPTUALISING POWER

The earliest studies of power (Marx and Engels, 1845; Parsons, 1960; Weber, 1968) considered how it is sourced, resourced and distributed. These analyses considered particular power bases within society, and there is a rich body of relatively more recent literature following in this tradition (Giddons, 1968; Mann, 1986). In the conventional sense, power refers to capacity to bring about change. Locke considers power as relational, being the capacity to make or receive any change (Locke, 1979). Lukes (2005) adds that it is also about being able to resist change.

Thus the basis of power frequently remains obscure as it is often only evident through its effects. But the effect of power is not the only means of inferring its existence (Barnes, 1988). Barnes argues that while individuals may have the capacity to exercise power, they may choose not to do so. Nonetheless they remain powerful because they are perceived to be so. This is played out through simple attendance at meetings. A successful business leader, with a strong network of connections, remains powerful because he or she is perceived by others to have stature in the community. Even though that individual may not choose to overtly use his or her position, the business leader is able to influence the actions of others by simply being present. This affects the behaviour of others who, in their efforts to secure some, uncertain future co-operation, may be anxious to comply with what they understand as that individual's beliefs, values and desires.

While the existence of power is not always very visible, the consequences of power are often observable and so are much more subject to definition. This challenge in itself may help explain why much of the focus of more recent debate is concerned with the exercise of power and the consequences of its effects and thus on how it might be identified, measured and categorised (see for example Mills, 1956; Dahl, 1961; Bachrach and Baratz, 1962; Lukes, 2005[2]).

The impetus for this study and the desire to understand the power dynamics of rural development has already been established. To succeed in this endeavour and to overcome the challenge of limiting the focus to action alone, it is necessary to consider more than the exercise of power. The source of power will offer essential insight into our analysis.

For pragmatic and political reasons much of rural development is measured in terms of its effects and influence on rural communities—this is an extremely straightforward and obvious way to evaluate the impact of programme activity. From a political perspective it provides a strong message to the electorate on the proficiency of elected representatives. Policymakers and politicians can demonstrate the efficacy and transparency of public spending, while also revealing how money is invested at a very local level and according to the desires of the relevant community. Deliberating on rural development activity in this way focuses on how power relations are executed, but it does not necessarily consider actors' source of power, their capacity to act (or indeed to not act) or the legitimacy with which they act, nor does it pay attention to their 'real interests' (Lukes 2005:28). It does not take account of less obvious power struggles that might reside beyond basic action, that is, within our wider understanding of rural development practice such as the pursuit of power games among actors or within the realm of policymaking.

Focusing on tangible or observable activity fails to consider the more subtle aspects of power such as that exerted by actors who choose not to act. This type of omission, Scott (2001) suggests, is critical. He claims that having power means not having to act and having the capacity to make a choice about one's actions on a particular occasion. Take the example of a community group that is planning a demonstration to protest against the inevitable closure of a local hospital. The group member who fails to verbally articulate support or indeed who actively votes against a demonstration, even though all other members of the group are forcefully supportive of the action, is indicating a position of power. That individual may then express her opinion that she believes it is not a good use of the group's time and would prefer them to consider high level lobbying with decision makers. This is a position of power as she realises the hostility that she will face when making her views known to the rest of the group. She also knows how symbolic the hospital is for many of the other members and as such her attitude runs counter to that of the majority. For whatever reasons, she does not feel that she must hold the same opinion as the majority; indeed she is empowered to challenge their beliefs outright.

Power may be less obvious. A local entrepreneur may induce deference within a meeting through his or her perceived status in the community. As a result community meetings may become less participatory due to coercion from certain members. Sycophantic reactions may also be evident as contentious issues are avoided or issues, which would under other circumstances be subject to debate, are agreed. Past altercations with a generous ally may result in a group acting in a way that is not necessarily a reflection of that group's mission, but instead echoes the interests of a powerful ally or of influential members. In so doing the less powerful individuals inadvertently shift their value-base to become more aligned with those who hold more power. Crucially they do not perceive this as a conflict, but are persuaded by the circumstances. And so we see how the power of the powerful operates across multiple contexts and in relation to many issues; they are effective without active intervention (Lukes, 2005). Consequently we need to 'search behind appearances for the hidden, least visible forms of power' (Lukes, 2005:86). Such analyses allow consideration of Lukes' fuller dimensions of power: as capacity to act (source of power) and the right to act (exercise of power and its legitimacy).

Lukes' original analysis implied the importance of structural matters in the analysis of power while focusing on behavioural matters; he made these matters overt when revisiting power more recently. Isaac (1987) explicitly calls for recognition that behaviour occurs within a structured context and power ought to be conceived in structural terms. This is one of many critiques of the so-called three-faces debate that has evolved over recent decades. Indeed in problematising the notion of real interests, Ron (2008) emphasises the dynamic relationship between social structures and social relations.

It is apparent then that we need to examine sources of power in addition to the place at which power is exercised and manifest to fully consider power relations within the broader rural development structures.

Power as a Resource: Distributive and Collective Power

Analysis of power falls into numerous categories. Among the different classifications that exist is the structural approach to power and resource theories encompassing distributive and collective power. It is these categories and their respective distinctions that are of interest here.

Weber studied the social relationship of actors in society; within this he determined power as a crucial element. He defined it as the following: 'Power is the probability that one actor in a social relationship will be in a position to carry out his own will despite resistance, regardless of the basis on which this probability rests' (1947:152). The Weberian power relation describes how one individual in a social relationship has *the capacity* to exercise power over another. Crucially it is a zero sum game, so that in order for one individual to gain power, another must experience a loss in power. It is a micro concept in that it is concerned with individual relationships. Weber's conceptualisation focuses on how power is distributed [among actors] within society and thus represents a distributive approach. Unlike Marx, Weber did not place economic power as the most important factor. Individuals do not always seek power for economic gain; often the status of the power itself is what they pursue. For Weber (1968) then, politics and culture or ideology are also sources of power. Elected government representatives that operate at the very local level do not seek power from the remuneration that they receive. Instead they gain power due to the status afforded them by their culture and so they are powerful due to their capacity to further political aims to which they subscribe. In this way power is quite simply limited and this understanding of power is commonly held within society today. If A becomes more powerful within a group then B experiences a loss of power. Executives in senior management teams are aware of the balance of power where if one executive performs excellently then he is rewarded with a large financial bonus, but correspondingly another individual's bonus will be reduced. The same principle applies for community groups that are applying to funding bodies for limited pots of money. There is therefore merit in applying this fundamental understanding of power relations to rural development. However it is limited as it does not conceive of the ability of the group to produce and thus increase power overall.

Parsons extended the power relationship beyond individuals to provide a macro perspective to the distributive concept. For him social action or interaction is a system that responds to other interdependent conditions. Parsons showed that 'power is a generalised facility or resource in the society. It has to be divided or allocated, but it also has to be produced and it has collective as well as distributive functions' (Parsons, 1960:220–221). Power according to this analysis is not a zero sum game. Parsons argued that through a system of co-operation, such as social obligations or sanctions, people can enhance their collective power. Viewing power through this expanded lens, he conceived of it as a generalised capability, as something that is endemic to social life rather than being tied to specific relationships. But this was more in the context of sovereignty rather than within a broader treatise of power. Hence, his fundamental idea was that the consent of an authority's subjects provides it with the capacity, or legitimacy, to act.

There are some potential difficulties with this paradigm. It implies a context of consent that legitimises an agent's capacity to exert power. By presupposing the establishment of collective goals the framework ignores the possibility of conflict or of a 'negotiated order' arising from differential powerholders (Giddens, 1968:265). It presumes the existence of an authoritarian and yet consensual relation between the powerful and less powerful so that everyone has the same aspirations. But historically community development activity is brought about in direct opposition to the state; it challenges the structures and norms that pervade state institutions. By its very nature, opposition and discord prevail. Meanwhile contemporary rural development groups operate within a voluntary system where on the one hand individuals have obligations to fulfil if they wish to be involved in that system. On the other hand they are able to make a choice and completely opt out if they do not concur with the group's mission. So for example they may not personally relate to the issue that is being addressed (see Chapters 6 and 7 for further discussion). Even if they do remain within the group it is unlikely that full consensus will ensue. Not all practitioners share common objectives, it is more likely that common ground must be negotiated. These difficulties aside, Parsons' analysis provides a crucial dimension to the debate. He usefully shifts the focus from the individual or group level and so draws attention to the significance of social systems and collective action. By working together groups are able to exert more power than if constituent members operated alone. Moreover, Parsons emphasises the importance of capacity, legitimacy and co-operation.

Concerns with where individuals, groups and finally, systems source their power all represent resource theories of power. Even though Mills (1956) and Parsons (1960) argued about the division between these approaches, they need not be seen as mutually exclusive (Mann, 1986; Heiskala, 2001). Mann defined power very simply: 'Power is the ability to pursue and attain goals through mastery of one's environment' (Mann, 1986:6). He develops the definition by demonstrating the relation between the distributive and collective aspects of power, thereby combining the traditions of Weber and Parsons. Mann claims that power relates to the power of one person over another but also to collective action; through co-operation people can increase their combined power over others and so the two elements co-exist. Co-operation and consent may arise as a result of the identification of common ground and compromised agreement rather than from authoritarianism. This type of collective power offers a context within which individual power is distributed. The latter becomes a sub-set of the former and thereby offers an augmented analytical stance that does not presuppose the notion of authoritarian relations. This is more akin to the characteristics of rural development, where individuals and groups frequently choose whether or not to participate in a collective. Then when they do participate, and as we shall see throughout this book, they engage in power relations at an individual level.

The co-existence of positive individual, group and system relations is crucial within rural development. Individuals may set aside personal differences so that the group is able to mobilise power to influence wider policy decisions. They may lobby the established system of central and local government on a range of issues such as planning and housing legislation that affect rural residents' ability to obtain affordable housing in their locality. Many groups establish or subscribe to umbrella organisations that serve to fulfil an influential role that individual groups alone are unable to achieve. Meanwhile within the group, people will embark on individual or micro-power relations. Some actors strive to become elected to official positions such as that of the chairperson, while others will apply pressure behind the scenes, attempting to sway the collective position of the group in a way that is favourable to them. Relating to the issue of planning and housing, a landowner may try to influence a group's attitude on a contentious proposed luxury housing development if he or she is set to make financial gain from the scheme, even though it runs counter to his or her more official lobbying position that discourages further development. Collective and distributive power thus intertwine and co-exist. This framework presents a useful platform from which to consider power dimensions in rural development. It is this broad meaning of power that I use in this book.

Power is about individuals and groups having the means and the ability to achieve goals that further their interests, all in the context of a larger social system. The goals may be to exert change or to maintain the status quo. However as in many studies of power, it is suggestive of change (Barnes,

1988; Shortall, 1990; Lukes, 2005). The focus of change here are the actions taken through processes of rural development and the ultimate objectives of those activities. Nonetheless the challenge remains, just as it did when Lukes published his original work on power (1974), how to identify the source of power. In other words where does power come from and how will we recognise it? How do we know when 'real' interests are subverted?

Identifying Power

Seminal power studies conducted in the 1950s emerged as a critique of American democracy. These debates originally focused on behavioural aspects of power. An elite theorist, Hunter (1953) claimed that the distribution of power did not tally with the popular concept of democracy. These theorists argued that power at both national and local levels was concentrated in the hands of elites such as business leaders and key figures in government who were neither accountable nor responsible in a typical democratic way (Hunter, 1953; Mills, 1956). In other words the minority elite powerholders do what they wish without due account to the majority electorate. Dahl attempted to show that power was neither as concentrated nor as irresponsible as the elite theorists maintained. He was a pluralist and while he maintained that the distribution of power was unequal, he argued that different actors and different interest groups prevail in different areas. Consequently power is distributed pluralistically and so there is no overall 'ruling elite' (Dahl, 1961). Common to both elite and pluralist theories is that the possession of power can only be identified in cases of overt conflict. Only concrete and observable behaviour, focusing in particular on decision making, is considered. If this were the case a ruling elite would only be identifiable if there was clear evidence that the (supposed) elite were able to exercise their wishes, even against majority resistance. The limitations of these theories have been well documented (see for instance Lukes, 2005); they do not consider the broader context within which decisions are made, nor do they pay attention to power that does not involve conflict. This constitutes a one-dimensional view of power that focuses on the public face of power.

A second dimension of power was put forward by Bachrach and Baratz (1962) who argue that Dahl's theory fails to appreciate the full dynamics of decision making. Pluralist analysis, they argue, only focuses on one face of power (the public face). They claim that power also has a private face that is evident in covert exclusion of the interests of particular groups. It is exercised through control or manipulation of the agenda, so that the very scope of decision making is confined to particular issues. They argue that this covert use of power makes possible the sympathetic and unproblematic public representation of power as serving the general interest. It is precisely this which means that they meet such little opposition. Both decision making and non-decision making must be studied with attention given to issues that are included and excluded, and also to the circumstances in which these events occur. By providing a context for the manipulation of decision making, the second face of power recognises the role of politics within social relations. But in common with the pluralists, advocates of the two-dimensional model consider only cases where overt or covert conflict exists, remaining focused on actual behaviour. If this conflict is absent then it is argued that consensus must prevail.

The third face of power was uncovered by Lukes in 1974 and revisited again in 2005. It has been the subject of energetic debates which have resulted in a deeper understanding of power in social life. Lukes claimed that, in limiting analysis to observable behaviour that encompassed conflict, whether overt or covert, the two-dimensional model did not go far enough. Power, according to Lukes, relies on the capacity and the legitimacy to call on the obedience of others. The three-dimensional model considers latent conflict which exists when there is a difference between the interests of those exercising power and the real interests of those they exclude. This third dimension of power is about being able to influence the thoughts and desires of the victims without their being aware of its effects; power is tied to agency. The powerless fail to recognise that their real interests are at risk and so make no attempt to defend those interests. And so Marx's claim that the ruling ideas are those of the ruling class is seen to hold true.

Luke's theory is based on the notion that the ideal of a community of autonomous persons, or individuals with agency; it is central to both the understanding and critique of modern society. Mindful of the preceding description of autonomous individuals and the fact that they operate within a particular context that is influenced by values, customs, norms and traditions, illegitimate power is an impediment to this ideal. It presents an obstacle to the achievement of individual autonomy in that one person can obstruct the actions of another. Power is seen as legitimate if it is based on the real or implied consent of an individual. Meanwhile exploits that influence those who are regarded as less than fully autonomous, and therefore without the capacity either to give or withhold their consent, are considered as illegitimate power. For instance a community group may be placed under pressure by local government officials to support a funding application to a specific regeneration scheme, unaware that by so doing they are removing the possibility of the allocation of funding and resources from that local authority to other activity within their community. In this case the members of the groups may be blithely unaware that their aspirations are being manipulated, that they are being subject to mild coercion and that their independence is obstructed. Without the ability for 'due reflection . . . it is hard to be persuaded that one is being dominated when one cannot come to see the chains' (Ron, 2008:11). Groups therefore need time to reflect on and discuss the issues that are important to them.

Despite this focus on individual and ultimately community autonomy, power is about more than analysis of its distribution amongst its members (Giddens, 1984). Significantly power is also a structural property of social life (Giddens, 1984; Lukes, 2005). Implicit within Lukes' theory is the importance of structural relationships within society, so that social relations confer power to social actors to act in certain ways. Power adheres to social systems as well as to individuals and groups within them; it relates to the capacity of agents to 'make a difference' and it is a structural property of society or the social community (Giddens, 1984:14). Social power plays a role in the deployment of resources and so Giddens (1984) argues that through actions, all agents have the capacity to draw on resources. Action is therefore seen as not only expressing the intentions of individual agents, but also serving to reproduce the structure in which such action occurs. Social structures are thus made real because of the action of individuals so that for instance the concept of networking is realised because individuals engage in this activity. In this way structure and agent co-exist as a duality. Relations, behaviour and structure become important.

Structures may be constraining to activity by discouraging particular action, but they also have an enabling role as they allow agents to do things in certain ways. Structures are 'the rules and resources upon which social agents draw when acting' (Porter, 1998:161), and they may be material and non-material (McGrath, 2001). By complying with rules of rural development programmes and availing of related funding, actors make choices and pursue strategies that are structured. Take the case of a group presenting a scheme to local government that seeks to regenerate its community. It will adopt a particular style that the group members believe reflects positively on them and demonstrates their ability to undertake the proposal. This is likely to be conducted formally and to use language that has resonance with the ideals of the government body. In this way Giddens (1984) embeds norms and values within the agent rather than in the structure. They exist virtually and so are only made real through meaningful actions; indeed they are produced and reproduced by action. 'A theory of power must analyse structural relations and the way they are worked out concretely by socially situated human beings' (Isaac, 1987:24). One cannot be considered without the other.

Structure and action are therefore interlinked, forming a circular relationship, a duality (Giddens, 1979). If this were not the case then community groups would be unlikely to engage in lobbying that aims to change aspects of social systems such as welfare benefits for those unable to work or establishing equal rights for same-sex couples. The labyrinth of possibilities for change expands and shrinks over time as actors and structures change and develop. This understanding brings centre stage the importance of institutions and social forces. But significantly it emphasises the role of the individual, this being central to our understanding and critique of modern society.

Exploring Individual Autonomy

The degree of independence of the individual and ultimately of the community requires more consideration if we are to study power in rural development practice. It is necessary therefore to consider Hayward's model of power and to do this we need to review Foucault's theories of power as it is from here that Hayward draws many of her ideas. Ultimately Foucault strives to identify the utopian ideal of social structures and institutions (Foucault, 1991; Rose, 1996; Murdoch and Ward, 1997; Dean, 1999; Lukes, 2005; Stanley et al., 2005). Illustrating the import of his seminal writings, many studies have applied Foucault's theories of power (see for example Flyvbjerg, 1998; Hayward; 1998; Raco and Imrie, 2000; Thompson, 2005). These studies illuminate our understanding of the way power works in practice, operating through institutions and resulting in conflict and struggle. They are however premised on the notion of the immanence of power which, as we will see, limits our ability to consider power relations between rural development actors.

Foucault asserts an ultra-radical position on power. He claims to reveal how power reaches into the everyday life of individuals through structural relationships, strategies and techniques through producing a micro-physics of power (Foucault, 1991). Of utmost importance, Foucault argues, are the structures and institutions. This is illustrated in the rhetoric of Foucault's seminal work on power and on the art of government. Here he is concerned with how government conducts itself to render its population governable (Dean, 1999). Focusing on how things should be, on archetypal models, Foucault emphasises design over concrete issues. With minimal concern for actual policies and the effect of the people involved, the individual position is marginalised. Indeed he does not believe in the autonomy of the individual whatsoever. 'The subject constitutes himself in an active fashion, by the practices of the self'; these practices are 'not something that the individual invents by himself' but 'patterns that he finds in the culture and which are proposed, suggested and imposed on him by his culture, his society and his social group' (Foucault, 1988:11). This denigrates the role of the individual in a social structure. It removes individual autonomy and rationality. If we are to consider the links between policy and practice then knowing what power looks like, who is engaged in its exercise and from where it is sourced becomes imperative. In fact this is critical given the premise that rural development operates within communities of rational individuals, all of whom have agency.

Over time Foucault develops his perspective of power. In the early discussions he offers a complex view in which governmental technologies are located between 'the games of power and the states of domination'

(Foucault, 1988:19). Where there is no possibility of resistance, Foucault argues, there can be no power relations. This is in contrast to the community debates on power, where real, implied or, as we shall see, coerced, consent must exist before power is evident. It is true that compliance with social norms and conventions does not automatically indicate the exercise of power (Said, 1986). In later work Foucault shifts the focus from the legitimacy of (governmental) power to consider the means whereby the *effects* of power are produced. He argues that power should be considered in terms of what it does rather than what it is. It 'must be analysed as something which circulates . . . It is never localised here or there, never in anybody's hands, never appropriated as a commodity or a piece of wealth. Power is employed and exercised through a net-like organisation . . . Individuals are the vehicles of power, not its points of application' (Foucault, 1980:98). Because power is dispersed in this way, it cannot in theory be linked to a specific organisation such as the state (Hall, 1980). For Foucault, and contrary to more common understanding (Hindess, 1996), power is immanent and is not something particular to an individual or an organisation: 'it is the name that one attributes to a complex strategical situation in a particular society' (Foucault, 1990:93). Foucault believed that 'power is everywhere; not just because it embraces everything, but because it comes from everywhere' (Foucault, 1990:92–93). But as Sayer points out 'just because causal powers are everywhere does not mean that they are everywhere equal' (2006:263). In any case this ubiquitous trait, and the emphasis on what power does, makes it difficult to study the source and exercise of power. Our investigation seeks to explicate the potentially uneven nature of power relations and thereby to understand inequalities within rural development.

Subsequent researchers have used empirical studies to develop Foucault's analysis. Hayward calls for the de-facing of power, insisting that 'students of power should focus on whether the social boundaries defining key practices and institutions produce entrenched differences in the field of what is possible for those they significantly affect' (1998:20). In a similar style to the community power debate, 'the subject matter of power de-faced, like power-with-a-face is power relations' (1998:footnote 27). However, she argues that power-with-a-face debates generated by Dahl's early writing avert attention away from questions relating to how power affects freedom and from critiquing social relations of domination by focusing on questions of distribution and individual choice. She claims that 'to exercise power is to act upon social limits to action: to act on legal, conventional and other social boundaries that define the field of what is possible for another or for the self' (1998:18). Hayward places structural issues central to analysis of power but unlike the power-with-a-face debate she denies that 'action is independently chosen and/or authentic' (2000:4). She claims that power operates impersonally, by 'shaping the field of the possible' (Hayward, 1998:12; 2000:118). Freedom is therefore dependent on an individual's capacity to act upon boundaries that constrain and enable social action, this capacity being promoted by specific social practices and institutions. But while Lukes embeds his analysis of the agent within a wider structure, the empowered individual is able to shape and control his life. He very clearly states 'that social life can only properly be understood as an interplay of power and structure, a web of possibilities for agents, whose nature is both active and structured, to make choices and pursue strategies within given limits, which in consequence expand and contract over time' (2005:69). So what is the extent of personal autonomy?

Hayward does concede that power de-faced is compatible with a belief in the value of the '*relative* autonomy' of human agency (2000:20, emphasis added). That notwithstanding, according to Hayward, individuals are so embedded in social structures that individual action can never be independent, in fact limits to autonomy 'are often institutionalised . . . not necessarily channelled through the actions of powerful agents who understand them or will benefit from them' (2000:34). She further asserts that because the way people act is in significant part an effect of social action, then it makes no sense to talk of 'free' action, much less to distinguish between free action and that shaped by the action of others. As a result Hayward calls for scrutiny of the boundaries defining the field of the possible rather than focusing on interaction, communication and other links between powerful and powerless actors. For us what is problematical within Hayward's analysis is her denial that humans are 'essentially agents with true desires, interests and wants, and/or the capacity to choose their ends and to act to attain them' (2000:21).

Hayward's de-facing of power has provoked debate and has contributed to our understanding of power and agency. Power de-faced seeks to consider asymmetrical relations through significant differences in social enablement and constraint among agents, highlighting inequalities in social structures. It urges researchers to move beyond individual power exchanges and instead to consider the boundaries defining social action and the effect of social practices and institutions within this process. With a focus on the politically relevant constraints to freedom, Hayward redirects attention from individual or micro-relations to structural matters. Her analysis focuses on how external constraints can impact on practice. For instance individuals participating in rural development in Ireland have little control over the development of rural policies within the European Community, even though they must comply with emergent programmes. So in relation to rural development, the constraints of spending particular funds as laid out by national government regulations, may restrict local groups undertaking particular project activity.

Many studies following from Foucault are lodged in social structures such as public institutions and policy. These structures are a critical component of rural development. They are however premised on the notion of the immanence of power which limits our ability to consider power relations between rural development actors. All the while they presuppose that power operates impersonally in a fluid manner. But the individual nevertheless plays an essential role within rural development (McAreavey, 2006, 2007; see also Chapter 6). Power has a face, individuals engage in power games. The distinction between free action and action shaped by others remains important to this study. We have already seen how actors are engaged in the practice of policy, typically through action that draws on values, beliefs and material resources. Individuals do not have access to the same resources, nor do they hold the same beliefs; as a result some are more powerful than others. And so while structural factors influence and frame such features, individuals retain an ideal of what constitutes their free action. As Spinoza pointed out 'men have always found that individuals were full of their own ideas, and that opinions varied as much as tastes' (Spinoza in Lukes, 2005:151). Much rural development policy is based on the ideal of the [relative] autonomy of individual members of a community. Individual freedom can become compromised through social structures, but also through the action of others.

Understanding power as something that exists within structures but is exercised by individuals provides a means of unpacking the relationship between structure and agent. It helps us to understand how actors utilise social relations, personal positions and the rural development framework to ensure the compliance of others to their positive end. It provides a mechanism for relating practice to policy and structure with agency. Social power is conceived as being 'distributed by the various enduring structural relationships in society and exercised by individuals and groups based on their location in a given structure' (Isaac, 1987:28).

Several critical questions that are pivotal to our investigation into rural development practice arise from these power debates: to what extent is there scope for individual agency to affect social limits to action? How much agency do individuals within rural development actually possess? To begin to answer this we will revisit Lukes' concept of power.

The Three Faces of Power: An Update

In his original thesis Lukes focuses on the exercise of power, and this has attracted criticism from other theorists (see for example, Isaac, 1987; Hayward, 1998) as well as from Lukes himself (2005). The analysis deals with asymmetrical relations, that is, with the power of some over others (and within this the securing of compliance to domination), and it only deals with binary relations between actors who are assumed to have unitary interests. Much of the focus of the analysis is on behavioural matters, with structural issues implied. Lukes contends that a fuller account needs to simplify the assumptions to address power among multiple actors with different interests and to consider power as a capacity, a 'potentiality that may never be actualised' (Lukes, 2005:69). He shows that even in a binary relationship, for example marriage, domination may characterise only some of the interactions and on some issues they may not be in conflict. On some matters individuals may be more powerful than on others. Lukes proposes that a better definition of power in social life than that given in his original treatise 'is in terms of agents' abilities to bring about significant effects, specifically by furthering their own interests and/or affecting the interests of others, whether positively or negatively' (2005:65). There may be multiple and conflicting interests within any one group as is commonly the case within rural development. As group members negotiate around these interests, power relations are revealed resulting in some individuals adapting their beliefs and values and altering their position. Analysis of rural development using the conceptual framework of power requires careful consideration of issues that are not immediately obvious. It must analyse how structural relations are negotiated by socially situated actors (Isaac, 1987) and so will enable a critique of both policy and practice.

Lukes provides an amended framework that reveals how the power of the powerful can be viewed as ranging across a variety of contexts and issues, encompassing intended and unintended consequences and being capable of being effective without active intervention. Those subject to power ultimately experience negative consequences, even though positive results can arise. This is an important point for our analysis and necessitates unpacking. The harmful consequences can be extremely subtle. It may render those subject to power as 'less free to live as their nature and judgement dictate' (Lukes, 2005:114), constrained to some degree from achieving complete fulfilment. It is here that we can begin to make the link to practice and relate back to our earlier discussion about rural development. If individuals within a community are not entirely free to act according to their inherent values and beliefs or with due regard to social, individual or material concerns, then their activity or that of the group is limited and practice is constrained. In this way power has been mobilised, individuals are unaware of their real interests and 'a community understands its own problems through frameworks that are in fact inadequate to address its own goals' (Ron, 2008:4).

So power is about more than straightforward domination or control as understood in common parlance with its connotation of coercion. Non-coercion may exist, positive results may ensue and all the time power has been exerted. It can be productive and 'compatible with dignity' (Lukes, 2005:109). It is also the case that members of the power elite can be sympathetic agents of given groups among policymakers (Mills, 1956:280; Hickey and Mohan, 2004). And so, depending on the range of options available to them and the consequences of pursuing those options, even those individuals who have options may be subject to power and domination if the options are loaded or constraints oppose their interests. There is no doubt that Lukes' radical concept of power is appropriate for critiquing rural development. With its attention on individual power relations in the context of wider social structures, it has the potential to inform our understanding of the relationship between policy and practice and ultimately of the efficacy of rural development. It is appropriate given its focus on the relationship between power and structure in social life, its attention on the individual and the ideal of a community of self-determining actors.

But a problem remains: how to identify these real interests? The framework provided by Ron (2008) is instructive. He advises that it is possible that individuals and/or communities understand their own problems through frameworks that are inadequate to address their own goals. Further, power relations affect public discourse and shape the way in which social actors understand problems. It may be the case that their real interests are masked through these processes. This framework has resonance with Mills' (1956) notion of the power elite determining the very structure of the institutions that shape the resulting social action and interaction. Finally, according to Ron, it is possible that real interests and domination 'can be seen as part of a dynamic process of moving back and forth between an understanding of the structure of power in society, and an understanding of the real interests at stake' (2008:4). So the conditions and way in which those interests are identified and defined are part of an ongoing and iterative process. Real interests are therefore likely to emerge as conditions change and individual perceptions, values and positions evolve, and as power relations are defined and redefined. Not only do we need to understand how the rural development process identifies and defines real interests, but we need to be clear about what we mean. It is possible that individuals are unable to articulate their real interests within particular social structures, they have no voice. This may be because they are not aware of the means of communicating their interests or they have no representation to do so. Equally those in power may not be listening. Real interests may also be obscured if individuals do not understand how to structure or frame particular social issues, challenges and problems. In this way real interests differ from preferences and wants (Ron, 2008). So while an individual may be able to express preferred options, all the while her real interests may be undermined if she is unaware of or unable to select a course of action not encompassed within the proffered choices. This has resonance with Arnstein's (1969) discussion about how information is presented: is it understood, does it make sense, do the 'listeners' have the skills to understand and question the matters that are discussed?

The dynamic process of uncovering real interests represents a complex relation between the power elite and the masses. This suggests that we need to critically examine the way in which rural development actors understand and identify problems within their community and subsequently the activities that they engage with to address these issues.

This can only be done through a close examination of processes of rural development.

The remaining chapters of this book examine power in the context of rural development policy and practice. This is achieved through a number of different dimensions—governance, participation and micro-politics. The linkages between social structures and social actors pervade the analysis, allowing us to analyse power relations and to review the connections between rural development policy and practice as highlighted at the outset of this book.

5   Micro-Politics Uncovered

This chapter positions micro-politics within rural development practice. Concerned with the 'intangibles' that bind groups together, it relates to trust; power; and personal attributes such as perceptions and motivations. It will demonstrate how understanding micro-politics is pivotal to gaining a deeper understanding of the interests of actors in the rural development process.

A detailed analysis of the theories of social capital and of micro-politics reveals their characteristics. Micro-political processes are shown to have many similarities to social capital. There are however a number of distinct differences that are uncovered in the course of this analysis.

WHY MICRO-POLITICS?

Micro-politics arises because of the reliance of rural development on the interaction of a number of individuals each of whom has his or her own personality, traits, and values and acts in different ways. Interaction is most likely to be face-to-face through public meetings, open sessions or casual conversation. It is therefore crucial that individuals are able to communicate effectively with one another. De Souza Briggs (1998) warns of the danger of meetings struggling along at needlessly high levels of confusion, distrust and resentment when effective understanding of, and response to, face-to-face encounters does not exist.

The significance of micro-politics to this research cannot be overstated. Power is about being able to influence the thoughts and desires of the victims without their being aware of its effects; it is a 'potentiality that may never be actualised' (Lukes, 2005:69). Even those individuals who have options may be subject to power and domination if the options are loaded or constraints oppose their interests. The 'power elite' derive their power from institutions, but also from 'their personal and official relations with one another' (Mills, 1956: 278). Consequently we need to 'search behind appearances for the hidden, least visible forms of power' (Lukes, 2005:86). To understand power we need to consider individual relations within practices of rural regeneration. However the nature of the micro-relations that emerge among these new partners of governance has to date only been given some recognition (Murtagh, 2001; Barnes et al., 2003; Scott, 2004; McAreavey, 2006) but has not been fully examined.

Few involved in rural development would argue against the importance of having effective meetings, positive consultation or useful participation. Yet, rural development/regeneration literature typically *implies* the importance of micro-politics with little outright or explicit reference to the significance of these intangibles and subtleties. One of the difficulties with exploring and analysing micro-politics lies in its very elusiveness. A group of people come together to achieve specific things, but what actually happens? How do they go about running their meetings? How do people interact? Given similar funding and development opportunities what is it that makes one process more effective than another? These questions can be answered through analysis of the micro-politics of rural development practice.

The identification and analysis of micro-political processes contribute to debates on community/rural development theory and practice in a number of ways. Forming part of the theoretical underpinning of social capital, this research serves to advance the social capital debate seeking to overcome some of the ambiguity that scholars have associated with the concept (Foley and Edwards, 1997; Portes, 1998; Fine, 2001c; Anderson and Bell, 2003; Shortall, 2008). Secondly, much previous and emerging research considers relations between the state and structures of rural governance, typically that of rural partnerships (see for example Edwards et al., 2001; MacKinnon, 2002; Thompson, 2005); fewer studies consider the relations that emerge at an operational level between rural development actors. Even those analyses that do examine relations between actors suggest that further knowledge of these relations would contribute to our understanding of the structures and processes of rural governance (Edwards 1998; Storey, 1999; Hayward et al., 2004; Shortall, 2004). By examining micro-politics we will have better knowledge of the practices associated with rural governance at the individual and group level. The significance of this research will have increasing relevance as the rise in popularity of the governance and partnership approach continues on a global scale (Cheverett 1999; Goodwin 2003; Gaventa, 2004; Shortall 2004; World Bank, 2004; Mowbray, 2005) with accompanying implicit reliance on positive group interaction and unequivocal emphasis on social capital (Portes, 1998; Woolcock and Narayan, 2000; OECD, 2001b; Grootaert and van Bastelaer, 2002; Fine, 2003). In sum an illumination of micro-politics can only serve to increase our understanding of what constitutes effective rural development practice, whilst also furthering theoretical knowledge.

RURAL DEVELOPMENT AND MICRO-POLITICS

The importance of micro-politics and micro-processes is starting to be recognised and labelled within the literature. Barnes et al. highlight the importance of 'micro processes' (2003:397) in constructing notions of representation and legitimate participation, calling for analysis of 'micro-politics' of interactions rather than sweeping statements (2003:396–397); Murtagh (2001) and Scott (2004) each investigate micro-processes within partnership structures. Meanwhile Taylor's research discovered that members from the community and voluntary sector felt that that they were simply involved in the micro-politics as they were 'working within rules that determine 90 per cent of how it's got to happen' (2003:191), citing the fact that the community simply has control over few things. While discussing planning issues, Lowry et al. voice concern about the way that group processes are 'sometimes designed and conducted in ways that—intentionally and unintentionally—limit participation and manipulate consent' (1997:178). They go on to call for greater attention to explicit and implicit group processes to help guide those working in the planning field.

Little is yet known about how micro-politics emerge in rural development practice, why it varies so much between groups and why it plays such a crucial role in rural development practice. This may be in part due to the very elusive and intangible nature of micro-politics, which in turn militates against distinguishing and subsequently analysing the concept.

Micro-politics is typically difficult to pin down. It can generally be described as the unintended subtle or intangible aspects of rural development that emerge through face-to-face meetings. Typically relating to informal relations, micro-politics is characterised by comments such as the following:

> '[Tom] just took over our last meeting. You couldn't get a word in edgeways. If he does it again I'm definitely not going back.' (Great Villham Action Plan group member, 21.10.99)

> 'We all knew [Rick] would be voted into the chair, it's hardly a democratic system.' (Community Project board member, 12.09.00)

> 'This project has nothing to do with our community, it's [Sue's] retirement project, not to mention ego-trip.' (Growthville resident, 20.08.01)

> 'There was a real buzz about the place.' (Market Town community champion, 20.01.01)

Inappropriate meeting behaviour, meetings outside meetings, bad mouthing the group and snide remarks further symbolise negative micro-politics. New friendships, new groups, the 'feel good' factor and positive social interaction are more positive aspects of micro-politics. Poor group relations can lead to spats and outright conflict while strong groups benefit from positive relations and negotiated compromise. In short both positive and negative aspects relate to what happens in the *process* of achieving broader rural development (social, economic or environmental) goals. It relates to the factors that are not instantly observable—the surface must be scratched and investigated to reveal the micro-politics at play. It is similar to the 'structures and processes beyond what is immediately perceivable' as described by McDowell (1992:213) on the subject of elite interviewing. It helps to make the story behind every good or bad project; this is not necessarily the one that funders or policymakers hear or wish to hear about, but it is often more revealing than an annual report or list of achievements. This is the 'glue' of rural development and regeneration. It is not always visible or obvious and is infrequently measured by funders but it is an untold (positive and negative) consequence of community participation and involvement: it occurs when individuals interact and can be described as the micro-politics of rural development.

There is no doubt that the process and practice of development programmes raises a raft of difficulties. Getting a grasp on rural development group minutiae alone can be complex. Rural development is presented for 'the community' and emphasis is placed on the ideology behind the community coming together. Grandiose claims are often made by policymakers of potential programme achievements. But as the previous two chapters demonstrate, and as Taylor (2003) argues, the rules of the game require a steep learning curve and heavy workload which rules many people out. In reality regeneration projects latch onto 'stars' (Taylor, 2003:194) and so the practice tends to involve a few key individuals who take on the role of volunteer or champion within their community:

> 'Where does [Jim] get his motivation from? This is definitely something special. I suggested that he might want to delegate a little more, but he feels that he has done very little anyway—in fact he is running the show . . .' (Research journal, 27.06.01).

> '[Growthville] again tonight . . . what's going on here? The full steering group should've been there tonight but it seemed to be the hardcore of councillors with the rest having dropped away' (Research journal, 08.03.01).

Thriving and healthy groups develop meaningful inter-personal relationships and actively enjoy the process of coming together. New friendships emerge or unexpected activities spin out from the group. These groups have a definite 'feel good' factor.

'We're in this together, even if we don't get this [SRB] funding, then at least we have achieved this . . . look we're sitting in the same room as the council and we're discussing the future of our community. Who would've thought it possible?' (Community champion, 20.03.00).

Basic interaction between practitioners in the regeneration sector demonstrates that those immersed in rural development implicitly understand the role of micro-politics. For instance the practitioner will grasp the importance of holding a meeting in what is perceived to be a neutral venue, or in any case rotating the venue to avoid a particular interest group dominating the process.

'The council chamber with microphones, fixed seating and wooden panelling was described as 'intimidating' even to those who were involved in many different types of meetings. The setting was staid, formal, overawing and suggestive of a bygone era. It is difficult to imagine how trusting relations encompassing reciprocity, moral obligations or even commitment might be achieved in such physically imposing surroundings. Ultimately it was agreed that meetings would be held in a number of locations including local schools, the volunteer centre, a sheltered housing scheme and a business centre' (Research journal, 13.10.99).

This proved to be a successful strategy in that each venue attracted different attendees and resulted in a variety of individuals putting forward their viewpoints. Had the partnership only met in a single venue, it is debatable if the same level of participation would have been achieved. Further if that venue had been the council chamber the image projected would have been contrary to a community-based partnership.

Those directly involved in the group appreciate the need for spending time on clarifying objectives:

'[The Village] project is very strong where many other regeneration initiatives fall down—the long lead in time that was taken with the process of constructing the actual questionnaire, mobilising volunteers and the time taken to process the results has evidently been worthwhile with the payoff being experienced now. A 90% response rate is great along with the action that is coming out of the appraisal—extension of doctors' appointments and the new gym facilities' (Research journal, 15.11.00).

Ultimately these complications affect the achievements arising from initiatives which are more typically modest, risk averse and anodyne than grand, innovative and radical. Time spent on creating positive micro-politics does not transparently contribute to the achievement of goals. Even so sensitivity to group procedures and processes, including covert and overt practices, would appear to be essential for generating funds. Government rhetoric and the discourse of development programmes reveals the importance of and a strong reliance on these intangibles with suggestions of equity, trust, empowerment, community, cohesion, networks and social capital in programme documentation (see for example DETR, 1999b; European Commission, 2004; Mowbray, 2005). Specifically funding may not be given unless a community has demonstrated that it has paid attention to these process issues that are affected by micro-politics (see for example participation in the Single Regeneration Budget, DETR, 1999b). Nonetheless traditionally micro-politics represents intangibles for which there is no direct support.

This overview reveals the intricate, sensitive and changing nature of micro-politics. Not revealed by the instantly obvious, it necessitates examination by close scrutiny of interactions resulting from individual positions, motivations, perceptions, personal characteristics and social relations, all occurring within a larger structural context. The group relations and norms that develop are affected by micro-political processes. They include power; trust; personal attributes such as perceptions and motivations; and legitimacy. While none of these are new, it is their unique combination and interaction that form the dynamic concept of micro-politics. As this chapter has already indicated, by its very nature micro-politics embodies intangible and elusive processes and so identifying, documenting and managing them is a complex and challenging task.

Although practitioners acknowledge the significance of micro-politics, it remains a relatively under-theorised concept in academic literature. With their intangible nature and focus on group relations, many of these characteristics have been aligned to social capital. We now turn to a critical analysis of social capital, including consideration of the relationship between micro-politics and this contested theory.

SOCIAL CAPITAL AND MICRO-POLITICS: THE SAME, ONLY DIFFERENT?

'Social capital refers to the institutions, relationships, and norms that shape the quality and quantity of a society's social interactions . . . Social capital is not just the sum of the institutions which underpin a society—it is the glue that holds them together' (The World Bank, http://go.worldbank.org/K4LUMW43B, last accessed 01.07.08). Social capital as a concept with its 'gargantuan appetite' (Fine, 2001a:12) has found popularity among many academics and policymakers across a number of dimensions, from the individual to the community, and across the globe so that is recognised internationally (see for instance Putnam, 1993; Falk and Kilpatrick, 2000; Portes, 2000; Shucksmith, 2000; Svendsen and Svendsen, 2000). The general fungibility of social capital to an array of contexts from time to discipline to place is highly problematical causing its likening to a 'conceptual monster' (Fine, 2003:587). As a catch-all, residual category, social capital becomes something of a 'black hole in the astronomy of social science' (Montgomery, 2000:228). Social capital is versatile, and that versatility has contributed to its widespread use both in academic research and policy development today. Complete books have analysed its application across the disciplines (see for instance Baron et al., 2000). Politicians employ the term: former US President Bill Clinton famously found inspiration in it for his State of the Union address in 1995 (Portes, 1998); meanwhile an Australian State Minister revealed the central importance of generating positive social capital for modern governments (Mowbray, 2005).

Interpretations of social capital abound and differ among academic theorists and across the continents. In North America most researchers associate the term with the explanation provided by Putnam (1993)—that is, of large collections of individuals or groups, often the community, with an emphasis on co-operation leading to integration and solidarity (Wall et al., 1998). Meanwhile many European interpretations tend to highlight differential power relations and social hierarchy (Wall et al., 1998), following on from Bourdieu (1986). Other writers follow in the tradition of Coleman (1988), connecting economic rationality to social action. It is therefore crucial that a clear description of the meaning given to social capital and to micro-politics in this book is outlined. In so doing it is hoped that the identification of micro-politics will provide some clarity to concepts that contribute to the creation of social capital.

Trust

The importance of face-to-face encounters and their associated intangibles is not always appreciated, causing them to go unnoticed or unmanaged with untold consequences. Bloomfield et al. (2001) state that trust is still most easily engendered by regular face-to-face discussions over an extended period. Trust itself is crucial, saturates much social action and has been assigned a central role in the creation of healthy societies. Indeed 'social life without trust would be intolerable and most likely, quite impossible' (Newton, 2001:202). But while the importance of trust is acknowledged, it is a highly challenging concept that is rarely the subject of investigation today (Gambetta, 1988; Luhmann, 1988; Newton, 2001).

Quite clearly, the concept and significance of trust have origins in Durkheim's (1984) theory of social integration; in Simmel's (1950) analysis of social exchange and reciprocity; and in de Tocqueville's (1969) emphasis on the importance of trust to building society. Many recent analyses of trust have been conducted in the context of healthy societies and social capital. It is often seen as the starting point of voluntary association and Putnam (1993) argues that along with involvement and co-operation, trust is an essential ingredient of networks of affiliation which in turn form crucial ingredients of society. Besides ability to compete, a nation's well-being has been described as being dependent on the level of trust in the society (Fukuyama, 1995). And of course trust has been described as the main component of social capital (Coleman, 1988; Putnam, 1993, 2000; Fukuyama, 1995, 2002). In a similar vein to Putnam, Fukuyama constructs an instrumentalist argument around the role of trust [and social capital] in civic society in the creation of economic institutions leading to economic stability[1]. But trust, like social capital, also has intrinsic value (Dasgupta, 1988); it is a valuable attribute for a healthy community or ultimately for a flourishing society.

Fukuyama provides a precise definition of trust that highlights its intangible nature and also reveals why it may be difficult to monitor, measure or even support. It is the 'expectation that arises within a community of regular, honest and co-operative behaviour, based on commonly shared norms, on the part of other members of the community . . . these communities do not require extensive contractual and legal regulation of their relations because prior moral consensus gives members of the group a basis for mutual trust' (1995:26). Not surprisingly then, trust encompasses an element of risk (Luhmann, 1980) with individuals making commitments on the basis of trusting that at some time in the future they will reap the benefits, either directly or indirectly. Trust is clearly related to social interaction and so will be evident as individuals engage in micro-politics.

Confusion exists between the notions of trust and of 'confidence' (Tonkiss and Passey, 1999:258). And so conditions can be created that lead to increased confidence within society. This is witnessed through increasing levels of regulation in the public and private spheres. For example the Sarbanes-Oxley Act of 2002 was created in the US as a direct result of the loss of public confidence in corporate accounting systems. Meanwhile in the UK, meeting the requirements of Charity Law remains an immense, and often overwhelming, task for voluntary and community sector organisations. The obligations are deemed necessary by government in the face of increasing incidences of irregularity in the sector. In this way concepts related to trust, such as confidence, can be managed through legislation, although it must be noted that they are costly to enforce. It remains the case however, that trust cannot be regulated and controlled by rules, but it relates to moral obligations, principles and reciprocal commitments[2] (Fukuyama, 1995; Tonkiss and Passey, 1999). What remains unclear is the extent to which confidence can be created as a precursor to trust.

As a component of micro-politics, I understand trust to operate at a personal level, albeit through social relations and structures. This takes account of family, institutional and societal factors and norms. And so while individuals may have confidence in systems, I focus here on the trust that they place in other individuals, albeit consequential of their belief in and expectations of broader structures. Trust has intrinsic value to rural development as well as being instrumental to achieving group objectives.

The Provenance of Social Capital

In the same way as the meaning of the term *social capital* has been hotly debated over recent years, so the origins of the term have been the subject of deliberations. There is debate on the exact date in which it was first coined; it has been assigned to various theorists. One of the earliest records appears in the work of Hanifan dating from 1916 (Putnam, 2000) and parallels with contemporary applications are evident:

> The tangible substances [that] count for most in the daily lives of people: namely good will, fellowship, sympathy, and social intercourse among the individuals and families who make up a social unit. . . . The individual is helpless socially, if left to himself. . . . The community as a whole will benefit by the cooperation of all its parts, while the individual will find in his associations the advantages of the help, the sympathy, and the fellowship of his neighbours. (Hanifan, 1916 in Putnam, 2000:19)

It was also used by Jacobs writing in the 1960s as well as by Loury whose work was published in the 1970s (Wall et al., 1998). Dube et al. (cited in Schuller et al., 2000) used the label in 1957 to describe public physical infrastructure in Canada, and by implication social infrastructure. Although its application is somewhat unlike contemporary usage, the Canadian authors do raise the means-ends debate in quite a raw theoretical argument. They maintain that given their contribution to civilisation, social capital and associated institutions are worth having for themselves, not just to facilitate other means (industrial development).

The academic underpinning of social capital is not new; indeed the concept has existed implicitly for a long time. Social and political thinkers have contributed over the centuries to the ideas that are currently applied to the concept[3]. I have identified two which are particularly germane to this analysis. Firstly the notion of positive participation in groups is not new, as this book illustrates. There are strong historical associations to the value of participation associated with the works of Durkheim (1984) and Marx (1964) and relating to overcoming anomie and alienation respectively. Secondly the importance of trust as a critical component of social life has been at the heart of intellectual enquiry for a long time. It was developed over the centuries by an array of social and political theorists such as de Tocqueville (1969) who noted the importance of trust in building modern society; or Tönnies (1955) who explored the differences between the organic conception of society, that is, community, or *Gemeinschaft* and the social-contract conception of society *Gesellschaft*. Later theories moved on from the dichotomous debate of traditional versus modern society to consider civil society and political culture (see for instance Fukuyama, 1995; Giddens, 1998). Trust is still viewed as 'one of the most important synthetic forces within society' (Simmel, 1950:326) and underpins both social capital and micro-politics.

Micro-Politics, Not Social Capital

Micro-politics is broadly defined as the intangibles occurring within a group as a result of the interaction of a set of individuals working together. It is about group relations and the norms associated with them. It relates to shared knowledge, perceptions, understanding, social networks, trust, values and traits that exist among group members. In some of the literature these nebulous concepts have been aligned to and described as social capital (see for instance Dhesi, 2000) and thus micro-political processes underpin the formation of social capital. Nonetheless, micro-politics is distinguished from the broader 'social capital' label in this chapter for several reasons. The concept of social capital has become so versatile that its value as a concept and tool for analysis has been rendered questionable. Secondly with an emphasis on positive outcomes from social capital there is a danger that studies are less analytical and more moralistic statements. Finally specific and subtle processes within rural development practice are highlighted and these, while contributing to the sum total of social capital, do not represent the entirety of the concept. To clarify these matters and to fully untangle micro-politics and social capital detailed analysis of both concepts now follows.

SOCIAL CAPITAL—BOURDIEU, COLEMAN AND PUTNAM

Social capital draws on a rich heritage of social theory, channelling a plethora of ideas into a single concept. It is hardly surprising that it has gained so much popularity and has been established as a cure-all for the ailments of modern society. In current usage social capital pertains to the ability to obtain resources through membership of or participation in social networks or structures. It relates to trust, norms and relationships that facilitate this action. There is broad agreement within the literature that the development of the current application of the term lies with French sociologist Pierre Bourdieu and American sociologist James Coleman. Robert Putnam is assigned responsibility for projecting it fully into the limelight, resulting in its widespread use in political discourse and consequent application in policy and academic circles. This escalation to popular discourse mainly derives from his interpretation of social capital as a community rather than individual benefit and of the instrumental function of social capital in creating economic value, matters to which we will return.

Originally writing in the context of the sociology of education, Bourdieu introduced an array of capital such as linguistic, economic and scholastic, later refining this to three: economic, cultural and social. 'Social capital is the aggregate of the actual or potential resources which are linked to . . . membership in a group' and providing credit to its members (Bourdieu, 1986:248). The amount of social capital that an individual possesses is dependent on the size of the network available to that person and the respective capital reserves owned by each of its members. He considered that 'the profits which accrue from membership in a group are the basis of the solidarity which makes them possible' (Bourdieu, 1986:249). Individuals invest and reinvest in the creation of networks and groups to ensure the institutionalisation of group relations in order to access the benefits of membership, that is, profits. This is an 'endless effort' that is 'necessary in order to produce and reproduce lasting, useful relationships that can secure material or symbolic profits' (1986:249). And so the three forms of capital are the principal fields that comprise a person's social position. In this way Bourdieu focuses on benefits at the level of the individual by virtue of group membership.

The work of Bourdieu focused on the fungibility of different forms of capital and this has certainly contributed to its ongoing attractiveness. While 'economic capital is at the root of all other types of capital' (1986:252), social capital Bourdieu argued, contributes to its creation. So individuals enjoy economic benefits such as investment tips through social capital. But the processes associated with investment in social capital do not have the same transparency as economic transactions. Obligations may be highly personal, such as feelings of gratitude, respect, friendship or institutionally guaranteed through the creation of rights. Consequently Bourdieu notes the ambiguity and intangibility associated with social capital, for instance highlighting the importance of unspecified obligations, uncertainty of time scales and of reciprocal expectations. This contributes to its concealment as a creator of other forms of capital, especially economic and is characteristic of social capital, keeping it distinct from the more formal economic market.

The second major figure in the modern conception of social capital is Coleman. Interestingly he and Bourdieu collaborated together on a conference in 1991 (Schuller et al., 2000), although famously they never referenced one another's work (Portes, 1998). Coleman defines social capital according to its function: 'it is not a single entity but a variety of entities, with two elements in common: they all consist of some aspect of social structures, and they facilitate certain actions of actors—whether persons or corporate actors—within the structure' (1988:S98). He focused his analysis on dense social ties underlining the importance of closed networks for both elite and non-elite groups and for privileged and disadvantaged individuals. Individuals' actions are structured and in this way social relations result from access to particular resources. In a similar vein to Bourdieu's link between social and economic capital, Coleman sought to bridge economics and sociology by introducing rational action and integrating it with Granovetter's notion of embeddedness as a means of analysing social systems (see Granovetter, 1985). Drawing directly from Granovetter, Coleman underlines the importance of 'concrete personal relations, and networks or relations [i.e. embeddedness] . . . in generating trust . . . and in creating and enforcing norms' (1988:S97). He identifies three forms of social capital, and as in Bourdieu's analysis, these are also intangible. They are: obligations and expectations dependent on trustworthiness of the social environment; information capacity of social structures; and norms associated with sanctions for individuals. Social capital thus 'exists in the relations among persons' (Coleman, S100–S101) helping to create social norms and sanctions that facilitate co-operation.

Both Coleman and Bourdieu confer the benefits of social capital to the individual as the result of community or family ties. But the individual has different motivations for generating the capital and the benefits are not quite the same. Social capital for Bourdieu is an intentional outcome and a direct result of involvement in particular networks, all benefits accrue to actors. Actors invest in the creation of social capital in order to change (i.e. enhance) their position in a social structure. For Coleman it is a by-product whose generation is only captured in part by the individuals who generate it, while some of the benefits flow back to the community. Coleman considers forms of social capital have a public good and/or a private good aspect so that it does not necessarily 'benefit primarily the person or persons whose efforts would be necessary to bring them about, but benefit all those who are part of such a structure' (1988:S116). And it is this 'public goods' aspect 'that leads to under-investment in social capital' (1988:S119). This is because individuals generating social capital according to Coleman have some altruistic motivations (even if this is limited) and the community in question benefits from their actions without the need for public investment. These issues of community and of co-operation and their subsequent development have partly led to the meteoric rise of social capital in modern society. It is to this that we now turn.

Putnam injects the notion of community benefit into social capital debates. Social capital 'refers to features of social organisation, such as networks, norms, and trust, that facilitate coordination and cooperation for mutual benefit' (1993:2). He draws on a respectable heritage of social theorists such as de Tocqueville, Hume and Hobbes to provide a framework in which to ground his central argument. It goes something like this: rich and vibrant civil societies with large amounts of social capital such as dense ties and social networks along with high levels of participation and trust, form the foundations for the development of that society and community. Social capital in this way enables collective action and bolsters good government and economic progress. Like Bourdieu and Coleman before him, Putnam continues to emphasise the central importance of economic capital to modern society, retaining its attractiveness for modern government. Somewhat unlike the earlier sociological studies of the subject, Putnam argues that the benefits accruing from social capital do not flow back to the individual; instead they are the property of groups and so become a community resource. It is a combination of these latter points that have contributed to the meteoric rise of social capital into mainstay rhetoric.

The extent of Putnam's influence should not be underestimated, and social capital as used today often interprets the flow of benefits to the group and the community. The types of connections made have formed the basis for its categorisation so that bonding social capital refers to ties between individuals in similar situations such as family and close friends. Bridging social capital indicates weaker ties among similar individuals such as loose friendships. Meanwhile ties between dissimilar people relate to linking social capital and enable access to more resources than would have been available in the community (Woolcock, 2001: 13,14).

Social Capital and Its Troubles

Currently social capital crosses disciplinary boundaries (Wall et al., 1998; Fine, 2003); with applications in public health research (Kawachi et al., 1997); economic development and development studies (Woolcock, 1998); political theory (Newton, 2001); and in migration studies (Zetter et al., 2006). An array of understanding can be identified within contemporary rural development literature alone (Flora, 1998; Falk and Kilpatrick, 2000; Svendsen and Svendsen, 2000; Shortall, 2004; Lee et al., 2005).

Despite this plethora of applications, or possibly because of it, it remains an under-theorised and over-simplified concept failing to engage with deeper seated issues of power and inequality (Foley and Edwards, 1997; Edwards and Foley, 1997). Not surprisingly the debate itself has caused some provocation among writers (see for example Woolcock and Narayan, 2000; Fine, 2001b). The narcissism notwithstanding, such charges indicate the dire straits of the social capital debate today.

Social capital is evidently associated with an array of concepts including trust, norms, networks, reciprocity and obligations (Bourdieu, 1986; Coleman, 1988; Putnam, 1995) and this has contributed to its popularity and application within contemporary studies. Consequently social capital has 'become a hot topic among social scientists of late. . . . the term has been used so often to mean so many different things that it has become the equivalent of an empty container, readily filled with whatever meaning the user—or the listener or reader—brings to the conversation' (Servon 2003:13). Although Portes (2000) and Bridger and Luloff (2001) conclude that more work needs to be done before social capital is adopted as reliable public policy, its emergence in the public arena is evident. This adoption has not gone unnoticed and its comparison to a 'chimera' (Wall et al., 1998:301) would seem apt as it moves in 'mysterious ways' (Fine, 2003:591). It is applied indiscriminately (Woolcock, 1998) with 'increasingly diverse applications' (Portes, 1998:2) as it is perceived to provide a quick-fix solution to a raft of societal problems.

Portes (1998) points out that contemporary literature on social capital tends to focus on the positive consequences. For instance while Coleman (1988) notes in passing the potential harm of social capital, as well as value, to others, his analysis focuses on positive consequences. Equally Bourdieu (1986) expresses momentary concern for non-elites who may be unable to permeate the networks that provided access to social capital. But their fleeting concern with negative aspects fails to fully expound the harmful face of social capital. Many studies following from this continue to exclusively consider the benefits of social capital. At a policy level, the World Bank has been accused of failing to address negative aspects (Fine, 2003). More worryingly, accusations of policy failure (for example the neglect of gender and ethnicity) in the era of social capital have been made resulting in a policy rhetoric that does not match the practice (Fox and Gersham, 2000; Hewison, 2002). Finally, although Putnam recognises the anti-social consequences such as sectarianism, ethnocentrism and corruption, it is the positive benefits that remain in the limelight (Mayer, 2003).

The positive and versatile manner in which the term social capital is often used implies it is a panacea for many of the difficulties facing communities (see for instance Putnam, 1993) and this in turn has caused debate on whether social capital really is a cure-all for modern society's ailments (Maloney et al., 2000; Body-Gendrot and Gittell, 2003). Focusing only on the positive aspect of social capital ignores negative outcomes, such as the exclusion of outsiders and excessive claims on group members, power struggles and conflict or the destruction of certain 'problematic' groups (Portes 1998; Fine, 2003; Mayer 2003; Shortall, 2004). In particular this was a point of criticism directed towards Putnam's earliest works, to the extent that he named Chapter 22 of his book *Bowling Alone* 'The Dark Side of Social Capital' (2000:350). Subsequently he states 'bonding without bridging equals Bosnia' (Putnam and Goss, 2002:11–12).

Not only does social capital appear to have seduced political leaders as they strive to be perceived as intellectually informed, but more tangibly it appeals to their cash-strapped policymakers as it provides a framework in which nonmonetary features may contribute to the creation of economic capital (Portes, 1998). For example the World Bank employs it as a social rather than an economic face to fiscal adjustment (Fine, 2001c), and this is deemed enormously acceptable in the 21st century. 'Social capital refers to the norms and networks that enable collective action. Increasing evidence shows that social cohesion—social capital—is critical for poverty alleviation and sustainable human and economic development' (http://www1. worldbank.org/prem/poverty/scapital/home.htm, last accessed 26.07.07). But the social capital of groups in a developed world is different to that in the so-called developing world (Molyneaux, 2002). The way in which social capital is used interchangeably across a range of dimensions is problematic.

In combining the role of the individual in the creation of social capital, along with its role in economic development, Putnam manages to reposition responsibility for the apparent decline in 'civicness'. Implicitly blame is shifted from business and professional elites to the less privileged masses in society (Skocpol, 1996). The message being that if social actors get involved in local organisations and networks, their community will thrive, or to use Putnam's language, they will become rich because they are civic (1993:3). As Shortall (2008) contends, this ignores the potential role of government in stimulating social capital as is the case in rural development programmes. It also subtly shifts responsibility for supposed social decline from government to society more generally (Portes, 1998; Lowndes and Wilson, 2001). But more fundamentally perhaps, while it aids our knowledge of collaboration and co-operation, it 'distracts attention from how social and political conditions structure that associational life' (Mayer, 2003).

The concept is further popularised by the fact that it places value on social relationships in political discourse (Schuller et al., 2000). It can therefore be viewed as a rebound to the eras of Reaganism and Thatcherism with their emphasis on the individual and parallel dismissal of society; and also as an antidote to the excesses that are associated with the modern capitalist world.

Finally the static nature of social networks within social capital is problematic. Bonding social capital relates to 'links among people who are similar in ethnicity, age, social class, or whatever—and "bridging" social capital are links that cut across various lines of social cleavage' (Putnam, 2004:3). This definition suggests that the strength of the links is the same, but the identity of individuals involved results in a different type of link, either bridging or bonding. By definition then, the type of links within a group with bonding capital can never change unless membership is modified. However rural development is about an ongoing process of change, micro-politics takes account of the evolving nature of group relations. So when a group meets for the first time, links among members may be weak. These links will be strengthened (or weakened) as a result of embarking on the regeneration process; either way change will have occurred.

MICRO-POLITICS AND SOCIAL CAPITAL REVISITED

Micro-politics and social capital are closely linked. Micro-political processes underpin social capital and indeed have commonalties with the characteristics of social capital; these will become clearer as the discussion in the following chapter progresses. Consequently, some aspects of micro-politics contribute to the issues that are the focus of social capital debates. There are a number of differences which will now be spelled out to illustrate the relationship between them and to reveal why this study is not one of social capital per se, but why it does offer insight into this complicated concept.

Analyses of social capital are on the whole structured, in that they recognise that individual actions relate to broader systems. This is also true of micro-politics; it occurs within a broader rural development framework.

There are however differences within the structural approaches. Social capital fails to pay sufficient attention to state agency in that the masses are held responsible for its decline. But it is known that support for civil society can be viewed as a kind of political laissez-faire so that the role of the state in society is altered, with civil society becoming a substitute for some state functions (Anheier et al., 2001). Meanwhile micro-politics is cognizant of this and of the actions and influence of the state.

Micro-politics places emphasis on social action without giving exclusivity to economic development (although of course some programmes are entirely focused on economic development) so that participation and collective action may be the purpose of the activity. This is important if we are to understand how communities and individuals determine their real interests and thereby gain a better understanding of power relations. Meanwhile social capital adopts an instrumentalist argument placing emphasis on ultimately achieving economic development and of making links to civic engagement and between economics and sociology. In these respects social capital and micro-politics differ.

The third difference relates to the intangible nature of both concepts. Both exist in the relations among individuals and embrace many of the same themes: power, norms, trust, obligations and reciprocity. As the name suggests, micro-politics considers this at a micro level. So for example while Coleman views trustworthiness of the social environment, I consider trust at the level of personal relations, albeit that this may be derived from or influenced by confidence of the social environment.

As indicated earlier, yet another distinguishing feature is the importance of considering both positive and negative aspects within micro-politics. Applications of social capital have notoriously focused on positive qualities. Further the failure of social capital to recognise the progression of links within and between groups neglects to acknowledge the dynamic and fluid nature of the development process. Micro-political processes operate within an ever-changing environment. Levels of trust change as do group norms and power relations. This all has a bearing on the success of the group.

Finally, the benefits of social capital in many academic interpretations today, and within policy circles are seen to lie within the community. Unfolding throughout this book is the notion of an elite rural development community. The benefits of and risks from micro-politics therefore do not flow easily into a community, but they are directly experienced by this more limited grouping that comprise the groups and individuals involved in regeneration, with only limited benefits going to some individuals in the wider community. The status of groups that are possibly in opposition to the state but are certainly outside of defined regeneration processes, is less well defined. These matters will be revisited in the chapters following.

6 Micro-Politics
A Taste of the Action

This chapter draws extensively from the research to highlight micro-politics in action. Different aspects of group dynamics are critically analysed, along with their relevance to micro-politics. They are connected to norms, agendas, communication, group objectives, hidden interactions and legitimacy. As the chapter progresses it will become clear that each of these is affected by power; all the while the concepts embedded within social capital, namely trust, norms and reciprocity are evident.

NURTURING TRUSTING RELATIONS

Making time for the creation of norms is a vital part of the development process as it can help prevent ineffective meetings and squandering time due to a lack of trust. The importance of positive personal relationships within the rural development process cannot be underestimated:

'... any point of view that I made was swiftly ignored by [Carol] but particularly [Andy] ... This is a very tricky exercise and one that I think is probably related to the amount of time that I've been involved with [Growthville] and their appraisal. Some of these people I've only spoken to for the third time ...' (Research journal, 03.10.00). But then '... some of the messages have finally got thro' about the public meetings. It's looking likely to be mid-March with an informal session and a more structured meeting to follow. It's just a shame that I wasn't able to join the [Growthville] group much earlier to establish the rapport that I now have, it might've saved some of this time wasting ...' (Research journal, 23.01.01). Later '[Jack] phoned me to provide an update on a meeting held last night. This is quite a move in itself—the fact that [Jack] actually considers me useful enough to call just to have a debrief' (Research journal, 26.04.01).

Similarly, the fact that I had a strong friendship with a community champion in The Village meant that I was privy to decision making. It provided me with the power to influence the agenda and the direction of the group. I had access to the private realm:

> 'I haven't mentioned this to anyone else as I look upon you as my main guide and strength so don't want to organise meetings you can't attend. Would evening meetings make it difficult for you to attend?' (Personal correspondence from community champion, 25.04.01).

Indeed failure to establish trusting relations may have a devastating impact on the success of a particular scheme. In Market Town the community dived into a consultation process that revolved around a completed questionnaire analysing perceived needs and priorities. When the results emerged, the group was not clear about how it would actually achieve any of these desired changes. Not enough time was spent at the outset deliberating on, and discussing, the process in which they were about to embark. People did not have time to nurture trust and empathy and so were unable to enter into meaningful or mutual relationships. Subsequent meetings consisted of repeating discussions or revisiting decisions previously made and of personal abuses being exchanged between particular attendees (Interview with community champion, 21.06.00). Trust is clearly a fairly amorphous concept and Fox's (1974) contention that it is eroded by perceived inequalities was apparent:

> 'There are at least two factions in this community and they are each trying to undermine the other by setting up rival groups' (Community champion, 21.06.00).

> 'Some groups and quite a portion of the population do not feel good about where they live, there is a strong perception that the needs of a section of the community are ignored by the council' (Community champion, 21.06.00).

Any semblance of trust was eroded as the individuals who had initiated the consultation process were blamed for lack of progress—many group members believed local authorities and other agencies were responsible for undertaking the identified tasks and activities. In turn local authorities and other agencies believed that a joint or partnership approach to potential projects would be adopted. Failure to establish clear lines of responsibility, to nurture personal relationships and to establish group norms from the outset had damaged the foundations of the group. Eventually, one year following the execution of a community appraisal, a 'regeneration day' was facilitated by an external consultant whose fees were paid for by the Community Initiative. This involved 'about 30 residents . . . trying to get people to prioritise and think in realistic terms and identify benefits to the community' (Research journal, 20.01.01). The meeting was deemed a 'success' by the professionals who helped to set it up (Professional practitioner, 22.01.01). Even so the community was effectively at a standstill for a year.

In this way the rural development structure was utilised to positively influence group relations. This was particularly observable elsewhere in the research as the following research journal quotes highlight:

> '[Growthville] Community College tonight for Village Appraisal meeting. The group are very enthusiastic having approx. 13 individuals they normally meet every week. They are hoping to circulate questionnaire by the end of March, gave warm welcome to [the Community Initiative]. The incentive of funding seemed to form a huge attraction, which probably holds some lessons for the future' (Research journal, 29.02.00).

> 'The SRB [Single Regeneration Budget] partnership met tonight. Meeting was v. well attended and perhaps this is to be expected given an announcement of funding in an area that has otherwise been starved of funding' (Research journal, 24.08.00).

The prospect of funding served to accelerate the development of positive relations among individual members. In this way the rural development structure enhanced the confidence of participants, paving the way for the emergence of positive micro-politics, particularly trusting relations.

Creating an environment that is conducive to meaningful exchange is not always straightforward. It can be tricky; inevitably it takes time and this is something that many believe the rural group cannot afford. Community leaders are often impatient to 'get on' and achieve tangible results, while officers involved with development groups are under pressure to achieve personal targets linked to their organisation. And yet face-to-face exchanges are vital for a group's healthy development, as they allow trust to grow and links to be strengthened. Scott (1990) investigated social performance through face-to-face behaviour and domination. He argues that those actors experiencing domination will keep information hidden until they feel that it is socially safe to raise particular issues. He claims that the more threatening the power, the thicker the disguise used. Problems can arise if a person continually masks a concern within a group. It may be as fundamental as anxiety about the group's focus, as happened with one particular member, Susan, of the Great Villham appraisal group. Her involvement was crucial as she was a resident of the area; the involvement of local people was minimal with the majority being professionals. While it was rumoured that Susan had misgivings about the lack of progress, she never actually articulated this at meetings. She obviously did not feel comfortable enough with other group members to air her views. Ultimately Susan withdrew her support for the group in a fairly dramatic manner:

'Unfortunately the CAP [Community Action Plan] seems to be falling apart—one of the younger and more dynamic members of the group stormed out of the meeting, from what I can gather this was due to sheer frustration ... This is not really a surprise, none of the group members could cite their reason for doing the CAP. Whose agenda are they really working to I wonder? Is [Edward] pushing this exercise? Why is the CAP group not more tightly aligned to the SRB partnership?' (Research journal, 21.11.00).

She then failed to attend any further community appraisal meetings. When questioned about this she cited frustration with the focus on appraisal procedures at the expense of tangible project activity (Interview, 30.08.01). At some level the absence of shared norms meant that trust and links within the group were fragile. The lack of regular and co-operative behaviour meant that individuals did not take the risk of remaining involved, trusting that the other members understood the process and that they were all committed to the same goal. Had Susan been embedded in such a trusting group environment it is likely that she would have been more outspoken, group norms would have been re-negotiated and her involvement would not have been lost. In this way diversity of participation was lessened due to poor micro-politics emerging from lack of trust and weak bonds within the group. This also demonstrates the fact that although rural development actors operate within a structured environment, ultimately they retain autonomy, they make individual choices and they can opt out at any stage.

Shared Agendas?

The role of the agenda in the rural development process should not be overlooked; it is often the place where power imbalances are found. Bachrach and Baratz (1962) describe a situation where consciously or unconsciously barriers are created around raising divergences; one person is prevented from bringing forward issues whose resolution might be unfavourable to another. For this reason the issues that are suppressed as well as those that are addressed are important (Lowry et al., 1997). This section illustrates how the agenda can be ingeniously used to further particular interests.

A meeting was organised by the local authority about the priorities of Great Villham's SRB application. At this meeting the officers did not present the options as either allocating local authority resources to working on the bid or of finding an alternative use for these resources. A local authority officer described the purpose of a consultation meeting in Great Villham:

'The question we are discussing tonight relates to the contents of the SRB application that is to be submitted. This will not be an easy process but we can work together to identify priorities and to put in a shared bid' (Research journal, 14.10.99).

If it had been a truly all-encompassing participatory exercise of community regeneration, the question would have been phrased differently. It might have asked, 'Do people wish the Council to put resources into the SRB?' Effectively the question of whether or not an application would be made was not up for discussion—the scenario was presented as an 'opportunity not to be missed' (Research journal, 14.10.99). People were therefore given the opportunity to participate in a limited exercise, that of developing a bid to the SRB and the restrictions that this entailed. Hence a barrier was immediately created preventing people who did not agree with this use of local authority resources from entering the debate. Powerful individuals (namely the elite SRB working group) limited the scope of debate to issues that were comparatively harmless to their interests. Attendees were prevented from raising their viewpoint and so they appeared to agree with the decision to submit an application for funding to the SRB. These individuals may not have spoken up because they felt uncomfortable about questioning what appeared to be a consensus among the group and they did not have the self-belief to publicise their anxiety about the issue under debate. The meeting norm or model was one of consensus and harmony rather than of disagreement and dispute. Consensus prevailed while controversy and discussion were discouraged. Anyone questioning the purpose of the meeting was deemed a 'troublemaker' (Research journal, 14.10.99).

The consultation with the community therefore rubber-stamped a decision previously made by the local authority to make a submission to the SRB programme. The discussion was limited to the form of the bid rather than concentrating on whether or not a bid should be made in the first place. In confining decision making to particular issues, covert power was exercised (Bachrach and Baratz, 1962); real interests were not identified or discussed. As a result a sympathetic and unproblematic public representation that was viewed as serving the general interest followed and was met with little opposition. In actual fact power was exerted within restricted areas through informal networks where access was limited to social and political elites (Woods, 1998a, 1998b). These backstage spaces are located away from the formal decision-making processes, intervention from opponents or the constraints of regulation; hence alliances can be developed and decisions agreed. Such decisions and alliances are then consolidated during the formal meeting. This was the case with the working group. The wider forum was powerless as it approved a decision made by certain individuals rather than the community. Hidden power was evident as those outside control were not only omitted from the political process but they were denied entry (Lukes, 2005). Although they were able to become involved in the forum and strand meetings, they were not invited to join the working group and thereby to participate in the restricted area. The ability of the partnership to identify the 'real' interests of the community was therefore most likely curtailed as genuine options were not fully explored.

In fact the local government body shaped the very decision that was taken, failing to take account of the voices within the community. The name of the community was used as a vehicle for furthering its agenda. It is debatable whether or not this was in the overall interest of that area; it could perhaps be more appropriately viewed as a proxy for the community. However what is clear is that the option was not presented and so the interests of the community were not fully explored. Krebs (1997) has indicated that we can appreciate our own interests better by understanding that of others. Ultimately the local government officers would have been well placed to appreciate the interests of the community.

Questions are often asked about whose interests are represented in rural development activities (Buller and Wright, 1990; Shortall, 1994). A key figure in one project was a retired community development professional who had worked in rural development extension programmes in Africa and Asia. It was he who drove forward the group's agenda and he was interested in 'improving life in the village, enhancing vibrancy and vitality' (Community champion, 24.02.02). He described the early meetings:

> 'We held meetings in all of the villages to survey the perceived needs. This was a complete failure—few came and those who did were the usual suspects. They looked at issues from their own perspective rather than from the view of the community as a whole'. Then in recognition of the importance of establishing a shared agenda 'people need support to help themselves. Things should not be imposed from outside. Initiatives should start from where the people are at' (24.04.02).

Pertinent to this discussion is identifying a shared agenda. It would appear that even in circumstances where there is no ulterior motive, the challenge of promoting the benefits of involvement remain. Problems persist when the masses are invited into the agenda-setting arena. Setting aside the dynamics of participation which are discussed elsewhere (Chapters 6 and 7), members of the community do not necessarily perceive the gains from involvement and the associated social capital or positive micro-politics that flow as a result. They may also have an issue with the individual that is driving the development process; we explore this matter further in the following section.

Playing Games in Growthville: Personal, Community or Elite Interests?

The following account illustrates the role of personal politics and interests in the rural development process. It provides substantial detail through excerpts from the research journal. These have been included to provide 'thick description' and give the reader an insight into micro-politics in action. The description shows how, through an appraisal that culminated in a public meeting, individuals pursued particular interests in an effort to dominate the proceedings.

A member of the community action plan group, Jack informed me during one of numerous one-to-one conversations that the community ultimately wished to establish a primary care facility in the village. This weighed heavily on his mind.

'[Jack] contacted me today to let me know of meeting 13 April which is to distribute questionnaires. The only sticking point has been the doctors' surgery and how to include?'(Research journal, 29.03.00).

He led the group as it organised a public meeting during which the results from the village questionnaire were to be presented, discussed and actions prioritised. Among the issues identified was the lack of affordable housing, the rate of new housing development, access to primary health facilities and the potential for a skateboard area for young people. The public meeting should have been an opportunity for the group to gain new membership and support for the broad work of the action plan group; this was explicitly expressed by other members during the planning meeting.

Certain group members wished to invite a range of representatives from public and voluntary sector organisations to join a platform panel at the public meeting that was reviewing the findings of the appraisal. Part of the thinking behind this was to identify areas for multi-agency working. The group believed that the pressure of a public setting could provide an environment that would short circuit a longer process of persuading agencies to work together. Originally Jack was against the idea of having external agencies present at all; eventually he was persuaded that it might be worthwhile to have them in the audience with elected representatives only among the platform panel.

'A very strange discussion ensues—[Conor and Jack] feel that [Jim] the local MP [Member of Parliament] should not come to the public meeting as it is for locals only—PC [Parish Council] and other councillors are ok as they are elected representatives. Very bizarre as [Jim] is also an elected representative. However, some cynics in the group feel that the MP would attend just to 'score points' . . . not quite sure what this means. Odd approach to meeting as there is never an agenda (they seem to be desperate to keep it informal). [Jack] has bizarre style of chairing. Not necessarily managing the process/discussion. This group sees the results/findings as something that should be handed over to the PC [parish council] for it to take the lead' (Research journal, 31.10.00).

Mindful of the role of the professional and of the community champion and the way in which they mediate between the local community, practitioners, funders and policymakers, the group was fast becoming a mechanism to pursue personal interests. Group members had spent a lot of time and effort distributing and analysing questionnaires. Health was one of a number of issues emerging from the community questionnaire as an area of concern. Being an elected representative and wishing to be seen to bring about change in the area, Jack was involved in many different ongoing projects:

> 'Another meeting @ Growthville . . . Once again [Conor] went on and on about how people at the public meeting would have to make the choice between increased taxes and more services or no extra services and no change to taxes. Even when I explained until I was blue in the face about the role of self-help groups I don't think he got it. I also have a feeling that the school is going to hijack this whole event as they are exhibiting their plans for the new sixth form college & this may arouse more interest than anything else. This is incredibly frustrating. The community is still hell bent on NOT inviting the MP but giving invitations to local councillors. No agencies are to come along and I think [Conor] nearly had a fit when the rest of the group thought the press should definitely be there!!' (Research journal, 13.02.01).

> 'I had a chat with [Jack] today. He is now aware of the fears of the school taking over and also recognises the need to have another steering group meeting to bring the other members back into the planning process. Apparently I'm not the only one to voice these concerns' (Research journal, 28.02.01).

It was clear that the group had not fully discussed how to prioritise and take action on the findings of the appraisal. Different interests, including health and housing, were vying for attention:

> 'There has been little or no clarity with this group about its role from the outset as we had another discussion about who would move things forward . . . the group would prefer the action plan to be picked up by the PC. However they don't seem to realise the importance and urgency of getting volunteers. Otherwise nothing is going to happen as a consequence of this project' (Research journal, 26.03.01).

Then unbeknownst to the rest of the group and despite the decision that had already been agreed by the group, Jack invited health sector officials to join him on the platform along with another elected representative. By doing this he countermanded the group decision and sent a strong message to attendees that health issues were to be discussed as a matter of priority. He also implied that his was a position of power, in that he had the capacity to over-ride group decisions.

> 'The event happened at last, [Growthville] open meeting! It was actually very positive, as well as having its negative bits. Positive in the numbers

that showed up (50+) and although the vast majority (c.80%) were older people, there were some younger attendees who made a contribution to the meeting. [Jack] was either very nervous or not very positive about the whole thing as his presentation was not exactly 'upbeat'. Meanwhile [Ray], a local councillor, stood up and assured the residents of [Growthville] that they might be able to avoid having any houses built as most of the housing development was ear-marked for places like [X, Y and Z]. And [Ray] is supposed to be representing [Growthville] . . .' (Research journal, 31.03.01).

Even after the public meeting, the interests of the community were not clear. A mix of interests continued to dominate:

'[Jack] phoned me to provide an update on a meeting held last night. This is quite a move in itself—the fact that [Jack] actually considers me useful enough to call just to debrief. And now he and the group are planning some follow-up meetings around some of the themes highlighted in the appraisal. But he was also interested in finding out how [The Village] got its doctors' surgery (the cynic in me suggests that perhaps this was the real reason for his call)' (Research journal, 26.04.01).

'. . . Apparently the Scouts and the football groups had got wind of the meeting and decided to gatecrash. The Scouts have a chip on their shoulder and feel that they have been ignored over the past 30 years by the PC . . . it was difficult to get the spokesman to move on from this ancient history. Despite the fact that the purpose of the meeting was to explore how services could be improved for older teenagers only two other young lads (with an interest in skateboarding) had come to the meeting . . . a circular discussion ensued between the scouts and the two lads . . . Then the other councillor seemed to be suggesting that the skateboarders carry out (yet another) consultation . . . I could sense their frustration at the chaos of the meeting . . .' (Research journal, 29.06.01).

'I spoke to [Jack] re: [Growthville] & he's very overwhelmed with things in general. In fact he sounds really stressed out—the library, the pub and the local plan review are all preying on his mind. I suggested that he might want to get others to help a bit more and he said that no-one wanted to do it. This leads me to think that maybe he doesn't really want to relinquish control, after all this is his part-time job. He did agree that we (the steering group) should have a review meeting to reassess where we're at and what we want to do next' (Research journal. 10.09.01).

'. . . Part of the problem at this stage is the group doesn't know how to move forward. It all seems terribly intimidating. . .'(Research journal, 20.11.01).

It became clear that Jack's motives were less than straightforward; he exerted covert power by attempting to influence the issues that the local politician dealt with:

> 'As for [Growthville], I think progress can be made if we show the councillors involved that it is not down to them personally to do things, but that smaller action groups can make progress. [Jack] also playing games, he has suggested to [Jim] (the MP) that the GP [Health] issue is the only one without progress—I found this out via the Chief Executive of the PCG [Primary Care Group]' (Research journal, 28.11.01).

At one level the group operated formally, using minutes, agendas and other meeting accoutrements to function, but at another level it had an informal basis. This revolved around trust and commonly shared norms, where group members expected honest and co-operative behaviour. Links within the group were apparently strong. However, Jack exerted his power in an authoritarian way disregarding both formal and informal group norms and betraying trusting relations. His manoeuvres were entirely structured. He knew how the political system worked and understood the importance of getting a strong message across from the beginning. By using social relations and accessing resources, that is, his connection to the MP, Jack was significantly furthering his own interest, that is, the health centre scheme. At the same time he was affecting the interests of others as alternative projects were sidelined in favour of the health project. This illegitimate power meant that those subject to it were not necessarily aware of it and were also rendered less free to live as they might otherwise have done (Lukes, 2005).

Jack used his ongoing role as an elected official to exert power, operate as he did and for his actions to go unquestioned in the formal setting. It may have been the case that while he betrayed trust within the group, more broadly he doubtless believed that his actions were acceptable for the greater good of the village community. Whatever the rationale, by directing the public meeting in a very particular way, Jack ensured that there was little opportunity for deliberation through an open public debate, thereby curtailing reasoning and critical judgement (Estlund, 1997) and jeopardising the identification of real interests. Not surprisingly the public meeting focused on health concerns and there was no meaningful opportunity for attendees to explore issues and priorities or for them to engage in useful social relations. Individual expectations were not met by the events of the meeting. More critically for rural development, as Chapter 4 illustrates, an absence of real or implied consent eventually becomes an obstacle to the achievement of individual autonomy. And so the ideal of community is not realised as individuals not only opt out of formal processes, but more fundamentally they are unable to freely choose a course of action.

Communication

The rules of rural development can create specialist language. 'Partnership! What does that mean anyway? We don't use that word. That's a council word' (Community champion, 26.10.00). Such was a community representative's retort on the Council officer's suggestion to incorporate the word into the project title. In the midst of such strong feeling and after heated debate, the Board agreed to steer clear of the word *partnership* on the basis that, not only was it meaningless to the community, but that it typically was used by government. The Community Project Board wished to portray an image of an organisation rooted in the community and belonging to Great Villham. The label of 'project' was agreed over 'partnership'.

Undoubtedly then language is more than a vehicle for understanding. Bloomfield et al. (2001) suggest it is not passive but is part of the development process. They go on to argue that it is 'reflexively constructed in relation to the contributions made by other participants, emphasising interpretation, feedback and revision' (2001:503). So language builds on what has been said before, with the latest contributor adding his or her own understanding to discussions before adding a personal contribution. Social actors perform differently between different audiences (Goffman, 1959). And so people say different things depending on the audience at the time with various dynamics influencing language including institutional interests, ethnicity, gender, class and personality (de Souza Briggs, 1998). Within partnership meetings the contribution of each member differed according to his or her status in the rural development sector; furthermore individual contributions differed according to the group context. For instance one of the local authority officers made a conscious attempt to use language that was not riddled with jargon when contributing to Board meetings where community representatives were present (Council Officer, 25.02.02). This contrasted with his more formal style within smaller working meetings where group members consisted of professionals.

Equally the language used by a House Manager was very different at the Community Project steering group meetings from that used during House Committee meetings. In the latter circumstances he borrowed heavily from the rhetoric and of the Senior Management Team assuming a very formal style laden with technical terms. Conversely at steering meetings he assumed a relaxed and lay communication approach. Language subtleties are not always apparent to everyone and those attendees not clued into these intricacies might remain oblivious, taking the meeting at face value—they do not appreciate the politics and the undercurrents/subtexts that lie within the meeting. The personal style of the Community Project chair was a reflection of that individual's performance to the other members in an attempt to be taken seriously by the professional members of the group. His style within the community forum for the same project was quite different. It was more relaxed and more encouraging of general discussion, placing less emphasis on focused discussion with clear action points. Herbert claims that such 'subtle behavioural variations reveal a deep and sophisticated cultural knowledge . . . that cannot be unearthed without abiding familiarity with the group' (2000:557). The Community Project chair played out various roles in accordance to what he believed to be appropriate given the different circumstances, indicating a comprehension of the perceived needs of his audiences and of the different functions of each group, but also of the structure within which he operated.

The Private Realm

In reality, important decisions, such as the earlier example about whether or not to allocate local authority resources to an initiative, are taken away from the public arena. According to Woods (1998b) it is the elite who have privileged access to or control over particular resources necessary for the exercise of power. The ability of House to make connections and develop important relationships external to formal structures was noted in Chapter 3.

This was exemplified in the way in which the senior management team (SMT) cultivated a close relationship with senior staff at the then Rural Development Commission (RDC) as it advanced ideas for the Community Initiative. These connections went beyond bridging or bonding capital, but could be described as linking ties, so that connections were made between those with differential access to power and resources. In nurturing such links the SMT became familiar with the priorities of the RDC. When the time came to make a formal submission for funding, House was in a position to make a strong case and ultimately had a favourable outcome. This association was noted by another agency whose staff felt that if their ogranisation had submitted an application for the same project it would have been unsuccessful (Professional practitioner, 04.03.02).

The group that went on to steer the project was seen by many as one which did not actually possess much power given the networking that had been done away from official and formal procedures. Some partners felt that their involvement was a token attempt by House and the funders to be seen to be inclusive (Professional practitioner, 04.03.02). 'Terms of reference for the group would have been useful and would have helped validate the role of the steering group' (Steering group member, 22.01.02).

While the lack of links within the steering group did not impede the success of the project, external links secured funding. These links were exclusive and obscure. The private realm by very definition is not a visible space; to elucidate activity behind the scenes, a discerning approach is required. Events cannot always be accepted at face value and often curiosity is necessary:

'Following some further investigation I discovered that the [Commuterville] Village Hall group and the Project Association were somewhat at odds with one another. I understand that the Project Association was chosen at a public meeting as the village's Millennium project. Contrary to [Joe's] account to me this was not a unanimous decision. . .I felt that no-one was being entirely up front about this and so I had to piece together snippets of information (Research journal, 26.01.00).

'I spent an evening with the [Commuterville] Project Association in order to go through their draft application. There is most definitely a lot more going on than meets the eye, I do not think there is such wide support for their application as they would suggest. Also I am doubtful that they have undertaken such extensive consultation as they are suggesting. I did recommend that they broaden out the project in order to involve local people a little more; this may also inform people more accurately of the nature of the project and thus increase their support base' (Research journal, 17.05.00).

As we saw earlier in this chapter, individuals play games. Some of these are hidden from the public domain, but this is nonetheless where trusting relations are developed and powerful decisions are taken. Not everyone has access to the rural development policy cabal, nor do they know how to access important community networks. Despite formal structures, the private sphere remains significant to the process of rural development.

Developing Relations, Establishing Legitimacy

The SMT at House understood absolutely the importance of legitimacy; this has already been noted in Chapter 3. Not only was the organisation well connected with powerful policy elites, but publicly it was involved with activities that enhanced its position within the community. The team recognised the contentious nature of housing in rural areas. The bottom line for House was that as a housing association its role was to build houses and rent these to tenants; in other words it sought to continually expand its housing development programme. As we have already noted this was a complicated process (see Chapter 3) and relied on the identification of suitable sites and accompanying planning permission. But the SMT understood that in order to be in a position to do this it needed to have a positive profile within potential communities. It therefore supported local events through for example small amounts of funding for family fun day events or towards playpark areas (House SMT, 14.02.01).

> 'We can't just go into a rural area and build houses, it's all about re-lationships. We need to develop good relations with landowners, with the parish council and with local groups. If we want to build houses we need the support of the local community. They must be able to trust us' (House SMT, 27.02.02).

The circumstances in which people become involved in a group influences micro-politics as it will affect their perceived legitimacy and thus their capacity to influence, to persuade or to engage in mutual exchange. Lowndes (1999) suggests that there can be confusion around where different representatives draw their legitimacy if the various mandates (election, appointment, common experience, professional expertise, and leadership skills) are not mutually recognised. Representation therefore requires careful consideration of perceived and actual legitimacy and thus power and status of the selected representatives. Not only was I employed by a Housing Association but I was also co-ordinating a community project that was sponsored by government agencies. I had access to resources and also access to a network of contacts that existed beyond the community of Great Villham. I was embedded in a greater rural development structure that many of the local groups were unable to access. I used these connections to enhance my stature. Eventually I was given the semi-formal position of leading the community and neighbourhood strand—one of three within the Community Project. I was also invited to join a small working group for the project development. From this point I had no difficulty in achieving mutual communication with the local volunteer centre; hitherto this had been highly problematical with the co-ordinator failing to co-operate at even a basic level. This was most poignantly demonstrated by the invitation from the centre to brief their network on the emerging bid where the invite stated that 'the bid is an important opportunity for voluntary and community agencies to partici-pate in a multi-agency partnership that might offer access to additional funding' (14.04.00).

Croft and Beresford (1992) point out that involvement and empower-ment is not a zero sum game and so individuals do not have to 'lose out'. By recognising the legitimacy of my position, the volunteer centre co-ordinator paved the way for the development of a mutual relationship. Our relation-ship had been altered in a manner beneficial to us both; it was founded on personal trust and grounded in broader rural development structures. My membership of the working group and of the strand leadership described above was gained in exchange for my skills and experience in consulting with communities elsewhere. As a member of the different groups I was able to enhance activities, rather than pushing other groups and individuals out of the process, highlighting funding opportunities and providing useful information relating to the rural development sector.

Trading tactics is commonplace among community participants. One of the community representatives, James, who was involved in the periphery of the SRB project knew that he was an important link between the community and the District Council. The Council had in the past been accused of being remote from its community. James was an important asset for the Council as it developed the SRB project and also the bid for Beacon Status (a kudos award from central government reliant on partnership with and involvement of the community). It could not afford to lose him, a point that was not missed on James. Consequently he was able to make derogatory statements about the local authority during planning meetings. In other circumstances he may not have been able to speak so freely, but the Council wanted him to be part of a group presenting the case for Beacon Status to central government so that he could demonstrate the links to the community and reinforce the legitimacy of the Council.

I was conscious of the importance of developing working relations with key individuals as a means of securing my legitimacy:

> 'I met with [Nancy] for lunch today . . . she seemed much more enthusiastic and positive about the [Community] project after I provided additional details . . .' (Research journal, 02.03.00). And later I attended a seminar that was co-led by Nancy '. . . but more importantly, this was another opportunity to build some bridges with [Nancy] and try and get progress in [Market Town]' (Research journal, 11.09.00). Despite this '. . . once again it appears that [Nancy] has been playing silly games—I found out lots of new information that had not been forthcoming. However I made it clear that funding would only be offered to the [Market Town] group if certain conditions were met' (Research journal, 08.01.01).

While I understood the importance of direct working relations, I also realised the structured nature of this. By providing funding I was enhancing the legitimacy of my position. On the one hand individuals who understand and are confident with the validity of their position are likely to participate fully in a meeting environment or in the rural development process, using their position to full advantage. They have a deep understanding of the structures within which development groups operate, appreciating issues of politics, power, networks and funding. On the other hand people who do not have this insight and thus confidence tend to be more accepting of the general consensus: they are unable to question motivations or to understand the different positions of various organisations. Legitimacy is a structured and hidden dimension of power affecting behaviour and performance of members and eventually the micro-politics of a group.

Review

Micro-politics are displayed through various mechanisms within a group. They are evident through individuals engaging in power games, communicating with others and from general social interaction. Communication affects micro-politics; it can help to strengthen legitimacy by strengthening the impact of messages conveyed. Language in turn is used to convey information and ideas, but it is also used as part of the development process. Practical issues inevitably have an impact on group relations.

Micro-political processes affect the strength of group connections or bonds. These features affect group development as they can reinforce existing barriers and divisions. Nurturing and subscribing to shared group norms, values and objectives are vital for the creation of positive group relations including trust, mutual co-operation and understanding. Otherwise there is a real danger that the group becomes stuck, struggling with apparently simple issues or negotiating and re-negotiating priorities.

Individuals may use their positions to influence the types of issues addressed by the group. An individual's power exists through relations and associations with others, both within the group and beyond, and so perception and image are crucial. Power can also be derived from social structures, for example from social standing and regeneration funding. This power can be nurtured through links developed in a private sphere, out of the gaze of formal regeneration activity. Individuals may use structures to exert power and so guide action in a particular way for the pursuit of personal interests. This may erode co-operation, resulting in a lack of consensus. In turn mutual ties are weakened as individual expectations are thwarted or trust among group members is jeopardised. While this may limit the success of the group in terms of subverting its activities, it has more significant consequences: individuals can opt out, typically irrevocably.

CONCLUSIONS

Micro-politics has been identified as a nebulous component of group processes and for this reason it is difficult to theorise and to research. It arises due to groups of individuals interacting and working together on shared activity. Typically difficult to pin down and identify, micro-politics is not instantly observable and so close scrutiny of a group is required to identify the process.

This chapter analyses the complex nature of micro-politics showing how group dynamics should not always be taken at face value as they are affected by a number of factors. In this way communication, while useful as a tool for exchanging ideas and information, can contribute to the development process. Actors need to be able to understand communication subtleties as well as the nuances of social interaction to make a full contribution to the rural regeneration process; otherwise communication can reinforce barriers and divisions.

Trust, power, and legitimacy combine to underpin micro-politics, surfacing from this are group dynamics including norms, perceptions, mutual relations and positive and negative social interaction. The result is the emergence of a group with a prevailing positive or negative atmosphere. That is to say some groups give off good vibes, while others exude an unhappy air. When a 'feel good' factor predominates, group members enjoy the social aspect of interaction while also progressing to achieve rural development goals. That goal may be the act of coming together to oversee a community appraisal; in itself this creates networks of association, something that is appreciated for its intrinsic value.

Alternatively relations may be more frayed with group norms consisting of disputes, negative experiences and a general lack of progress. Micro-politics has some similarities to social capital in that both are structured and intangible and both draw on some of the same classical sociological concepts, namely trust and power. Understanding micro-politics will contribute to the knowledge base of aspects of social capital, but they are distinctive. They have negative as well as positive impacts and the flow of benefits back to the community is more limited than that of social capital. Whereas social capital denotes a range of fixed ties (bridging, bonding and linking), the mobilisation of which allegedly brings benefits to the greater good, this analysis reveals how micro-politics relates to fluid links both internal and external to the group that bring about more limited prospects. The positive, or sometimes, negative development of a group can ensue. That group is a restricted category, typically of a few key individuals, often operating as a proxy for the community. The wider implications for the community are yet to be considered.

Micro-politics is revealed through various mechanisms within a rural development group. It is displayed by individuals engaging in power games founded on trust and legitimacy. Individuals must be seen to have legitimacy, whether derived from their pre-existing position within the sector, or resourced through other means such as access to information and expertise. Trust is a consequence of informal relations; it is not something that can simply be created at will. Conditions must be appropriate to allow individuals to establish mutual relations that incorporate moral obligations or reciprocal commitments. Developing norms within the group and among its members advances the establishment of trust and the recognition of legitimacy. However external agents can contribute to this process, as was evident through the implementation and interpretation of regeneration rules and regulations. Therefore while micro-politics concerns group relations, these occur within the structured context that is rural development.

Power exists as a result of people acting together and it is also here that legitimacy is acknowledged. Relations and associations are more powerful than individuals and so perception and image are important aspects of power and legitimacy. Power tactics are exerted within group dynamics and affect meeting norms such as decision making and agenda setting. These were found to be complex, encompassing decisions that are made as well as those not made, how decisions are reached, which issues are discussed/which are not and the individuals influencing and controlling the agenda. Consequently the process of decision making and agenda setting can favour one rural development agent over another. By operating behind and beyond the group's agreed meeting format, backstage elites take decisions around potentially contentious or important issues. The less powerful group members feel unable to contest or question their action. Such hidden and subtle dimensions of power reinforce existing barriers, with the less powerful individuals remaining in a weak position. Nonetheless more positive tactics may be employed when individuals trade their positions in ways that are favourable to both parties. This can be done to achieve legitimacy and power within a group, resulting in benefits for the whole group such as access to information or networks. In this way power becomes a collective resource, rather than a zero sum game, and the benefits are felt beyond the individual level. Rural development agents perform according to the circumstances in which they find themselves and the image that they wish to project—their actions are to an extent structured.

It was stated earlier that micro-political processes, by their very nature, are elusive. Once identified and their causes understood, groups can take steps to manage them. Although very simple measures are often required, their successful implementation can be more complicated; they may require additional resources, extra time or may be met with opposition. For instance achieving a balance between formal and informal relations helps to ensure that both styles are catered for within meetings. This may entail using two types of meetings rather than a single one to further group objectives, so requiring additional resources. Equally challenging is the task of managing group relations through ground rules that outline group processes, as this can generate suspicion or erode positive components of micro-politics such as trust. While support from external agencies may help alleviate poor group relations, some group members may believe that such outside advice is not appropriate or necessary. The group's control may also be compromised through alliance with another agency. Furthermore using such measures may also represent the difference between an enjoyable and sociable experience and one which is laden with procedures and regulations.

Successful rural development relies on the positive interaction and dedication of typically small groups of individuals. Moreover people remain involved with initiatives because they enjoy the benefits of social interaction while achieving other specific and common goals. The process of rural development is therefore an end in itself, in the same way as many of the issues affecting micro-politics are valued intrinsically, that is, power and trust.

Individuals become disillusioned with rural development because of negative consequences such as personality clashes or abuses of individual power. These elusive social processes, positive and negative, are the micro-politics of rural development. From a policy point of view micro-politics is often an unintended consequence—it is not something that policymakers and funders can readily measure. Rural development rhetoric places less emphasis on micro-politics or group processes than on group objectives and outputs. However the reality is that groups are caught up with these matters, individuals are animated by micro-political processes and they are a vital part of rural development often making or breaking a process. All the while micro-politics reveals the intrinsic value of the process. Along with Chapter 5, this chapter offers a theoretical and empirical analysis of micro-politics, highlighting the role of micro-politics in the success of rural regeneration projects. However, as we stated at the outset, agents cannot be considered in the absence of broader social structures. It is to the structures of rural development that we now turn in Chapters 7 and 8.

7   Unravelling Participation

This chapter has a simple objective: it seeks to scrutinise the meaning of participation in rural development. It considers participation in the context of governance and development programmes. This includes an analysis of how participation is used to establish and support regeneration structures. A distinction is made between bottom-up and top-down approaches to regeneration and development.

Participation has influenced World Bank, European Union and many other development and regeneration initiatives. It tends to be used freely and glibly among policymakers and academics. Participation is one method used 'to inject some notion of the 'common good' into the functioning of governmental institutions' (Murdoch and Abram, 1998:41). Instinctively the concept of participation has positive connotations; it is considered a 'good thing' and yet the purpose of, and eventual gains from, participatory activities within rural development are not always clear. Hayward et al. point out there is a 'mythologising of the power of participatory methodologies to accomplish problem solving, emancipation or empowerment' (2004:95). Clearly the notion of participation is not unproblematic and as a result it 'generates enthusiasm and hostility in equal proportion' (Croft and Beresford, 1992:20).

In the 1960s Arnstein talked about understated euphemisms and exacerbated rhetoric around the concept, resulting in great controversy about the whole subject matter and so citizen participation is 'a little like eating spinach: no one is against it in principle because it is good for you' (Arnstein, 1969:216). The World Bank definition of participation describes it as 'a rich concept that means different things to different people in different settings' (The World Bank, 1997:11).

Nonetheless participation is embedded in current rural and local development programmes as they attempt to reconfigure regional structures of governance. These programmes depend on citizen participation in a community which is supported by socially inclusive structures. They emphasise the development of rural areas' capacity to support themselves through 'capacity building', 'community-based initiatives' and 'partnerships' (Buller, 2000; Ray, 2000; Shortall, 1994). The power conferred to the new partners from the private and civic realm is the subject of an emerging debate within rural studies (see for instance Shortall, 2005; Thompson, 2005; Goodwin, 2006). At both theoretical and policy levels, this model and its principles have generated debates about its effectiveness and ability to deliver on its promises. Given its connection to politics, power, social capital and social exclusion, it would seem that a critical examination of the complexities of participation will start to unravel some of this wider debate to reveal what it means to participate in rural development. That is the over-arching purpose of this analysis.

This chapter seeks to unravel the meaning(s) and dynamics of participation by examining theories of participation. Firstly an investigation of its theoretical underpinnings reveals its complex and multi-faceted nature, highlighting the importance of power relations to participatory practice. Then, using evidence from the empirical research, the chapter considers the role of the participatory process in rural development structures.

THE MAGIC OF PARTICIPATION

The popularity of participation is propelled by many different ideals from within society. Historically the dangers of non-participation were seen as sociologically significant with the advent of modern, industrial society; anomie, or social disaffection, was closely aligned with suicide rates (Durkheim, 1984). Participation is a perfect antidote to Marx's notion of alienation (Marx, 1959) whereby individuals in a capitalist society experience loss of control of the influencing powers within society. This alienation is experienced in relation to all major institutional spheres such as the political economy, religion or the economy. Although created by humankind, they appear alien to individuals within society. This is seen very clearly where many individuals do not work under their own direction and so have little control over their lives. Participation is one way of overcoming the deficit caused by anomie and alienation.

Participating in community is also seen as embodying direct relationships in contrast to the unfamiliar world of the state. Classical sociologists believed that community would disappear with industrial society (Park, 1952; Tönnies, 1955). Yet community is a concept that has a great deal of currency, and the search for roots, belonging and identity is of considerable sociological and political concern (Delanty, 2003). The idea of community embodies belonging, holistic understanding of needs, a collective desire to advance the good of the area and so on (Shortall, 1994; Shortall and Shuckschmith, 2001). The direct relationships couched within community are the result of participation by people because they feel a sense of belonging. They are also the consequence of inclusion and result in the creation of society that is socially, economically and democratically integrated (Commins, 2004).

Participation is viewed by policymakers and politicians as a constituent of the well-being of the human condition, but also essential to the quality of democracy. It thus remains a key indication of a healthy, engaged, and equal society. The reformation of political structures that have been central to new modes of governance highlights the importance of participation in modern social structures. While formal structures of government are prevalent, alongside this are the less formal arrangements of governance whereby the nation-state is one component within growing international and regional structures that are involved in new ways of making decisions (Giddens, 1998; Jessop, 1990, 2005; Hajer and Wagenaar, 2003). The belief is that decentralisation and participation make for better government. This is because decentralisation brings government spatially closer to people and increases the availability and quality of information from government to citizens. And so decentralisation enables citizens to more actively participate in structures of governance allowing greater involvement or social inclusion.

Politicians and policymakers are attracted to participation for further ideological and economic reasons. It is seen to generate social capital through which actors are able to secure benefits by virtue of membership in social networks or other structures (Bourdieu, 1986; Coleman, 1990; Putnam, 1993). It is the basic argument of Putnam's very influential work (Putnam, 1993, 2000). Putnam argues that the quality of society is compromised by non-participation and can lead to problems of exclusion. He also famously states 'Development economists take note: civic matters' (Putnam, 1993). The implication is that civic participation leads to economic development. This connection between sociological and economic perspectives appeals strongly to financially constrained policymakers (Portes, 1998). As a result it is used to justify participatory development programmes (Babajanian, 2005).

Finally participation has been shown to be closely connected to power (Arnstein 1969). Increasing citizen involvement can be seen as a way of minimising the role of the state in society by substituting many of its functions while also restricting state power (Anheier et al., 2001). It introduces the notion of shared responsibility giving individuals opportunities but also requiring them to accept obligations (Blair, 1994). Being seen to hand power over to 'the people' is a valuable political instrument. It 'adds a different and valuable dimension to local decision-making processes' (Stoker, 2005:12). It helps resolve tension between the state, market and community and it provides a 'softer' more people-centred approach than is otherwise possible (Adams and Hess, 2001:20). Meanwhile the softened rhetoric of community suppresses the visibility of the power of the state (Levitas, 2000). Consequently as responsibilities are shared, traditional public and voluntary sector boundaries become 'blurred' (Stoker, 1998:17), so that some activists would argue that participation is not about minimising the state, but rather increasing the responsiveness of political institutions (Anheier et al., 2001). The extent of power sharing in these participatory structures remains unclear.

Defining Participation

Participation is defined here as involvement in community activities that further the development and implementation of public policy. This definition takes account of formal and informal activity and includes direct and indirect benefits to public policy. By taking account of public policy, recognition is given to the role of structure in social life. Direct links to public policy might include a consultative exercise by a local authority on proposed housing developments. Other indirect links emerging from participatory practice might include the implementation of community action plans by local communities. However my definition does eliminate extreme political organisations such as the voluntary activity of groups with aims and objectives that are not in the public interest.

Participating is clearly a complex affair. The experience of participation may vary greatly depending on where it is happening; who is leading and involved in the process; the underlying function of the participatory exercise; and access to resources and expertise. Difficulties arise, tensions exist and conflict can ensue. Despite these mixed attitudes and with a few other exceptions (see for instance Shucksmith, 2000; Cleaver, 2001; Hayward et al., 2004; Shortall 2004), academic enquiry has tended to assume inherent benefits of participation. There has been less debate around the quality and legitimacy of non-participation or of power differentials and process issues within participatory approaches.

Participation and Its Woes

Hayward et al. develop the notion of non- or peripheral participation and challenge the assumption that 'broad based participation is always a social good' (2004:96). They implicitly show how participatory activities can lead to displacement, citing the example of the inability of a community to raise a football team due to the success of the local internet café. This suggests that communities have a saturation point for community-based activities and so full participation in one initiative may limit participation in another. Full participation is thus not necessarily an optimum position for community regeneration and so it is more appropriate to consider participation that is legitimate, relevant and inclusive.

Callanan reminds us that difficulties should not imply that participation is a 'bad thing' (2005:927). The existence and validity of non-participation should be recognised. Wenger (1999) argues that relations to communities of practice involve both participation and non-participation. Hence not everybody identifies with everything or everyone encountered and so it is normal for participation to consist of experiences of 'being in' and 'being out' (Wenger,

1999:165). Consequently not all members of a community will identify with a regeneration scheme for their area and participate accordingly. According to Hayward et al. (2004), while non-participation may indicate social exclusion, the act of non-participation does not imply social exclusion or lack of empowerment, but it may actually be an act of empowerment. Individuals, their research demonstrates, may choose not to participate and yet remain an active member of their community. Thus applying Wenger's principles to rural development, individuals who do not identify with particular activities exist on the outside of that boundary, in this case the rural regeneration boundary. Individuals may choose to opt out of participatory activities for various reasons such as consultation overload, lack of time, dislike for the participatory method selected or lack of interest in the particular theme of the initiative (Lowndes, 2001a; Hayward et al., 2004). In other words there are legitimate and valid reasons for non-participation often based on rational choice. This is in contrast to common rhetoric that designates it as undesirable.

Non-participation through choice is different to imposed non-participation. As Hayward et al. (2004) suggest, the act of non-participation may indicate exclusion. This occurs where barriers exist that prevent certain individuals and groups from participating to the extent that they would otherwise choose. Hence Wenger (1999) identifies marginal groups among those who participate less than fully, arguing that this occurs because full participation is *prevented* by a form of non-participation. In other words barriers to full participation prevail. For instance a community may be so used to the dominance of the local authority in shared activities that the norms of the local partnership reflect formal approaches of that local authority. This leads to comments and questions such as 'are we allowed to do this?' as the culture of non-participation is so ingrained among those on the margins.

Problems arise in marginalised participation because barriers exist to full participation. For instance if formal groups only are recognised by a partnership and thus given places on a committee, informal voluntary groups are marginalised. This is exemplified in Williams' (2003) critique of the UK government's Policy Action Team report. He highlights how on the one hand the document considers increased involvement as activity occurring through formal voluntary or statutory organisations, while on the other hand it discusses the development of informal activity to more structured action. Williams concludes that the report reinforces the notion of formalised groups as constituting more superior or legitimate participation than any activity that occurs outside this highly structured sector. He warns that, if the government continues to adopt this rigid approach, areas with more informal community involvement will lose out to those areas with traditions of highly formalised community participation. Barriers need to be considered, identified and overcome and this can only happen when the objective and purpose of participation are understood.

Less than full participation also occurs when some degree of non-participation is perceived in a positive way. Wenger describes this as peripherality and it can be identified when some non-participation is necessary to *enable* a kind of participation that is less than full (this is different from full blown non-participation which occurs when individuals actively and completely opt out). Thus in community regeneration, individuals may choose to contribute to the group organising a family fun day while opting out of the working group that is writing a funding application. Although their participation is not maximised in the sense that they are not involved in everything, they make a legitimate and relevant, albeit peripheral contribution. It is likely that many individuals will exist at varying degrees of the periphery as they participate in some activities, but not all. They choose 'to participate within self-defined limits' (Hayward et al., 2004:101).

If there are no barriers to full participation, non-participation or peripheral participation is not necessarily problematic. It becomes a problem when individuals are completely excluded or if they exist at the margins. Barnes et al. (2003) suggest that to examine participation more closely it is necessary to look at the notion of representation. Whose interests are being represented? It is from this position that I wish to contribute to the debate on participation and rural regeneration. The rhetoric of rural development would suggest that Mills' gloomy account of the power of the ordinary individual is inaccurate as rural development is about empowering regular people.

> The powers of ordinary men are circumscribed by the everyday worlds in which they live, yet even in these rounds of job, family, and neighborhood they often seem driven by forces they can neither understand nor govern. 'Great changes' are beyond their control, but affect their conduct and outlook none the less. (Mills, 1956:3)

This chapter will investigate the extent to which individuals are part of the rural development framework. This will take into account the broader context within which participation occurs, the nature of decision making and power relations between the stakeholders. It adds to the preceding discussion on micro-politics (see Chapters 5 and 6), by considering the rural development framework and the ability of individuals within a given community to participate and in turn to affect these structures.

POWER REVISITED

As we have already discovered, the notion of power is not unproblematic. Chapter 4 provides a detailed analysis of theories of power. There are however some additional points worth noting within this discussion.

We saw how community studies of power emerged as a critique of American democracy in the 1950s. Community debates on power show how social relations confer power to social actors to act in certain ways, thereby advocating the autonomy of the individual. These theories are based on the notion that the ideal of a community of autonomous persons is central to both the understanding and critique of modern society. This is appropriate given the premise that rural development operates within communities of autonomous and rational individuals. It has further significance in a system of governance which allegedly decentralises power outwards from government to the individual actors.

Original community power debates focus on behavioural aspects of the public face of power considering the powerful as 'those who are able to realise their will, even if others resist it' (Mills, 1956:9). While criticism has been made that the one-dimensional approach does not consider the broader context within which decisions are made, in fact attention is drawn to the relationship between structure and agency, so that a 'close understanding of the institutional landscape in which they act out their drama' is necessary to understand the power elite (Mills, 1956:280). This is somewhat in the manner described by Miliband (1969) where he contends that individuals within the ruling class have a determining role within society and are in a complex relationship with the structures of society; in other words they influence the very structures in which they exist. It is in contrast to other Marxist theorists such as Poulantzas (1969) who claim that the ruling class may participate in the state, but they do so as agents within objective structures rather than as actors in interpersonal relations. According to him, any overlap between the interest of the ruling elite and the state is by accident rather than design. To be clear: in this study power is understood as a dynamic and multi-dimensional concept; it pays attention to individuals' profiles, their relations with others and also to institutional mechanics giving rise to their position (Mills, 1956:280).

Exploits that influence those who are regarded as less than fully autonomous, and therefore without the capacity either to give or withhold their consent, are considered as illegitimate power. Power is seen as legitimate if it is based on the real or implied consent of an individual. Power is not necessarily a negative feature of social life, but ultimately those subject to power are constrained to some degree from achieving complete fulfilment. Their real interests are subverted by certain power relations; according to Ron we need to be able to move 'back and forth between an articulation of a view of real interests and an in-depth understanding of the power relations that distort them' (2008:11).

Given this meaning assigned to power and its connections to governance, it would seem that a critical examination of power in patterns and processes of participation within regeneration will contribute to existing knowledge. The significance of this research will have increasing relevance as the rise in popularity of the partnership approach continues on a global scale (Cheverett 1999; Goodwin 2003; Marinetto, 2003; European Commission, 2004; Shortall 2004; Mowbray, 2005; Bochel, 2006; Gilchrist, 2006).

CONFIGURING THE COMMUNITY PROJECT

Partnership, one of the key concepts in rural governance, was certainly at the heart of the Community Project. A fairly complicated structure emerged for this particular scheme with approximately 50 partners drawn from public, private and voluntary sectors including residents' groups, local authorities, housing associations, health trusts and business groups. Partners were committed to share the aims outlined in the Single Regeneration Budget (SRB) submission. The Community Project aimed to raise 'the Economic and Social Capacity of [Great Villham]' (SRB final bid). Activities included 'social and economic capacity building', achieving 'improvements for the quality of life of the residents' and 'enabling all residents to participate fully in society' (SRB final bid). The types of organisations that were funded included Age Concern, Sports Clubs, Citizens Advice Bureau, the Volunteer Centre, a toy library, youth projects and St. Johns Ambulance.

The local authority faced a dilemma: it wished to respond to the tight timetable of the SRB programme (three months from initial expression of interest to submission of final bid), while also engaging in a meaningful way with the local community. It considered that three months was insufficient time to engage with the number and variety of stakeholder interests in a meaningful way (Council Officer, 20.03.00).

A working group was established by council officers and comprised of two council officers along with representatives from a local business forum, a rural development agency, a health organisation and a housing association. The individuals were handpicked on the basis of a number of factors, as articulated to me by a council officer:

> 'We had some limited experience of consultation and public engagement. It was useful to have your input. Your contribution to the debate around neighbourhood renewal was helpful as well as your role in pushing and developing this' (Council Officer, 25.02.02).

This elite group took a pragmatic decision to use existing networks and connections to kickstart the consultation process on the basis that wider and deeper participation would follow. Prompted by this advice from the working group, the forum chairman stated in an early meeting that

> 'the partnership must not be exclusive and although 30 organisations have been invited to the meeting, there will be other important organisations who will hopefully wish to become involved and that one of the purposes of the discussion groups would be to identify other contributing partners' (Forum minutes 22.03.00).

An excerpt from an invitation letter, written by council officers regarding a strand meeting to discuss the bid encouraged the addressees:

'If you can identify any group that you feel should be invited please feel free to copy this letter to them' (28.03.00).

The fledgling partnership faced the 'revolving door syndrome' (Taylor, 2000:1020) where usual suspects, rather than disenfranchised groups, are primarily involved. It remained difficult to attract new individuals to the process:

'Despite a free buffet, not much interest was shown, with few new faces at the meeting' (Research journal, 11.04.00).

This desire for maximum participation notwithstanding, there was some evidence of pseudo-participation (Deshler and Sock, 1985) with the consultation meeting to discuss the contents of the bid (see Chapter 6 for further discussion).

The officers believed it was imperative that control ultimately remained with them, and they claimed that they could not 'simply put the project in the control of another organisation' (Council Officer, 23.08.00). In fact they proposed recruiting staff from their organisation to the post of the project manager.

The Council issued a paper 'Proposed Administration and Delivery Structure' (03.08.00). Discussing this in a Forum meeting, it was made clear that

'ultimate decision making powers lie with [the] District Council as they are the financially accountable body, but they will be advised by the Management Board, who will in turn be informed by the three Strands and the Consultation Forum'(24.08.00).

Council officers in the Community Project had an inflexible view of power relations. Implicitly they understood power in the Weberian tradition as a zero sum game: if they handed power over to community representatives, they believed their own control would be eroded. They did not conceive of power as something that could be tapped into and mobilised by the collective whole for the greater good (Parsons, 1960; Mann, 1986).

Consequently 'I met with the Council's solicitor to discuss the terms of reference for the Board and the Council. It is fairly clear that while this is a partnership by name, ultimately the Council are in control. Of course this means that they are accountable and responsible, but the structure is such that even if things go wrong, they will be able to implement a slick PR plan' (Research journal, 25.10.00). The possibility for the use of public relations brings to mind Mills' observations and the potential for belief in the message of the elite. He recognises that 'most American men of affairs have learned well the rhetoric of public relations, in some cases even to the point of using it when they are alone, and thus coming to believe it . . . Yet many who believe that there is no elite, or at any rate none of any consequence, rest their argument upon what men of affairs believe about themselves, or at least assert in public' (1956:5). Reading back through my research journal the complexity of the relationship between the local authority and the regeneration project is in no doubt:

> 'According to [Peter] their open culture within the organisation comes directly from management. I think he's right as [the senior team] is very open and approachable and all of the officers seem similar. A sharp contrast to other Local Authorities in the county' (05.02.02).

But on reflection I wonder how much of this is actually true, or is it true because Peter stated it to be the case! Quite clearly there are power differentials at play. An individual wishing to participate in the regeneration of the area had the option of doing so under the terms set out in the national SRB programme. This was not a 'popular space' that emerged from within and was defined by the community; it was an 'invited space' of governance in that it was an arena created and defined by government and was one into which communities were invited by the state (Cornwall, 2004). In the language of development programmes, it represents a combined top-down and bottom-up approach. On the surface this would seem to correlate to Lukes' understanding of power (2005:65). It relates to agents' (primarily the local authority) bringing about considerable effects by both furthering their own interests while in this case also affecting the interests of others (the individuals and groups that got involved in the Community Project). It remains unclear the extent (positively or negatively) to which the interests of others were affected; but what is crucial to this research is the fact that an elite group appeared to retain control of the regeneration parameters; the autonomy of individuals participating in this framework was eroded. This will be further analysed in the following chapter.

Establishing appropriate organisational structures was an important aspect of the project's development and one that was required by national programme guidance. The final Community Project structure comprised a Board, a Forum and three Strands addressing the following themes: community and neighbourhood; local economy and training; and health. Each of these structures is considered in turn.

The Board

The working group that was instrumental in writing the SRB application was subsumed into the Board with additional members being co-opted to provide representation from parish, town, district and county councils and from the three different strands (see later for strand functions). There were no elections as individuals were co-opted by the council officers.

Aside from the parish and town council representatives there were no individuals on the Board whose membership was based on the fact that they lived in the area; they were all associated with a particular organisation, agency or group.

Among its agreed objectives was that of negotiating 'the unfolding vision of the [SRB Project] with all partners' (Board minutes 12.09.00). Other aims were strategic development, monitoring and development and implementation of the SRB project delivery plan.

The Board's key function was to oversee the implementation of the SRB project overall, including financial and practical aspects. Getting the balance between these different objectives proved to be challenging:

> 'We [SRB Board] had a difficult dilemma over the process being pulled from a number of directions: being accountable, having systems in place whilst also getting on with the task in hand' (Research Journal 30.11.00).

Other key functions included appraising project applications and appointing staff. The possession of a range of specific and complex administrative and financial skills was required to participate as a member. More generally it was noted that there may be a 'lack of knowledge and experience of effective multi-partner delivery schemes' (council officer note to strand co-ordinators 14.07.00). It also instilled a certain degree of fear among local authority officers as the following research journal excerpt reveals:

> '[Council officers] were pretty honest in admitting that the bureaucracy of this programme is crippling' (26.09.00).

The expertise that was valued was that possessed by those writing the regeneration rules; it was not surprising that the membership consisted of professional practitioners. To paraphrase Mills (1956), the very framework of rural regeneration confined the rural development actors to projects that were not their own. Quite clearly with skill requirements as they were, and without a strategy of developing new talent, the participation of inexperienced individuals was always going to be limited. We will revisit this issue in the following chapter.

The Forum

The Forum was initially used to consult with individuals from the community over the proposed content of the bid and so provided a mechanism that allowed members of the public to participate. Eventually its function was to provide a mechanism whereby interested individuals could communicate with the project and vice versa. Generally approximately forty people attended these meetings. They were advertised by word-of-mouth, by placing posters in public areas—including shops, community centres and schools—and through direct invitations to agencies and community groups. A number of techniques were used within the meetings to engage with attendees including public discussion, focus groups and networking.

The meeting of 27.09.01 was themed 'Funding opportunities for voluntary organisations and community groups' (Forum minutes, 27.09.01), and its format illustrates a typical Forum meeting. It included informal presentations from regional umbrella organisations as well as from locally based voluntary organisations. Local groups involved in fundraising spoke about their experiences. The meeting received an update of the ongoing work of the Community Project and oversaw the election of two representatives from the voluntary sector to the Board. It began with tea and coffee at 5.30pm, with the formal business starting at 6pm. Following an open question and answer session, the meeting ended just after 7pm. On the surface the Forum appeared to advance wide participation.

Closer scrutiny reveals the controlled and limited nature of the Forum and of the overall project. To paraphrase Mills (1956), even though the residents of the area were living in a time of big decisions, they were not making any. Many of the attendees to the Forum consultation meetings expressed concern over the threatened closure of the local hospital. They clearly believed that this was the appropriate vehicle to raise the matter and did so during the question and answer slot. They failed to understand that Forum consultations were part of a distinct participatory process bounded by the SRB rules and to the terms defined by national government. A comment made by a council officer at a public consultation makes this clear:

> 'I know the threatened hospital closure is very important to everyone here. But I'm afraid we can't provide funding for that. What we can do is support the wide range of projects that are outlined in the bid document' (22.03.00).

Participation was an output of the regeneration programme, 'slotting' into pre-determined and externally defined aims (Oakley, 1991; Jones, 2003), rather than a process that also ascertained views and shaped the bid. If we apply Lukes' (2005) revised understanding of power in relation to agents' affecting positive or negative effects to advance their own interests, then these regeneration actors were not empowered. The content of the meeting was highly prescribed; there was little room to veer away from SRB activities and power was retained in the hands of the elite, in this case predominantly the members of the working group.

The Strands

Meanwhile three Strands were created encompassing the following themes: neighbourhood and community; health; and local economy and training. They were established to facilitate input from professionals and users of the Community Project. Their purpose is reflected in the invitation to the first community and neighbourhood strand meeting. Voluntary and community sector organisations were asked to participate to 'identify high priority needs' for their group and 'to ensure all relevant voluntary and community organisations are represented within the bid' (28.03.00). The selection of Strand co-ordinators reveals complex power relations.

As manager of a rural regeneration project I was one of two co-ordinators of the Community and Neighbourhood Strand. Strand co-ordinators were hand-picked by the Local Authority Officers involved in the project. The Officers were looking for individuals with whom they could work as well as people with useful skills and contacts:

> 'My background is in council led business programmes. Engaging with the community was quite novel for me and I went through a steep learning curve. I was glad to have [the Community Initiative] there to share consultation techniques' (Council Officer, 25.02.02).

The process favours those already in authority; this outcome is not surprising as Mills (1956) reminds us that the power elite are comprised of individuals who, through selection and training, are similar. The skills, attributes and qualities that are deemed useful are more likely to be held by like-minded individuals making up the regeneration elite as they value and possess similar skills. This is nurtured through their training; the exchange of 'good practice' evident in publications and by sharing ideas at regeneration practitioner conferences.

As a Strand co-ordinator I had strong views on my position. Reluctant to take on the role of co-chair in the first place as I believed it ought to be a position for a local community representative, I was persuaded by officers from within the council to do it at least in the short term:

> '[Council Officer] would like me to continue with co-ordinating community/social strand (alongside [the rural regeneration agency]). I am happy to do this ... I also suggested to [Council Officer] that the [Community Project] management group should have a 'voluntary representative' as it is purely made up of paid staff. Interesting to hear his comments, he does not want a lot of controversy and he considers that this might make things 'tricky' (research journal, 23.08.00).

As the project progressed and in the course of discussions with these officers, it became clear that this position was highly coveted by them. They viewed me as a safe choice, whereas an individual from within the community might be less predictable. Dahl (1961:226) outlines various sources of power in social relations including access to money, information, social standing, charisma and legitimacy. The importance of these sources was evident in this research. My position was valued in part due to my prominence within the regeneration structure. Specifically this was the 'prestige' (Mills, 1956:83) of my employer's standing and authority in the regeneration sector resulting from links with powerful funding and policy bodies (Community Initiative Evaluation, 25.02.02). But it also arose from the knowledge and contacts that I possessed and the fact that the officers knew that they could work with me personally (ibid).

> 'I kept the link with [The Community Initiative] because of you as an individual, your knowledge and experience and your contacts. Knowledge, guidance and the ability to signpost to other agencies was an invaluable contribution from [The Community Initiative]. This went hand-in-hand with you as an individual. If a different person had been running the project, we might not have had this link' (Council officer, 24.02.02)

> 'You have been greatly valued because you are independent. The time you have been able to commit has been invaluable. We will have to look hard to fill this gap' (Council officer, 28.02.02).

I underestimated the role of the personal in this process. Somewhat in contrast to the local authority officers, I had envisaged a process whereby someone might take up a position of deputy chair, to eventually become a chair/co-chair, the latter position granting an automatic place on the formal Board. In such a way true capacity building would occur, allowing individuals to develop and implement their skills via the project. This scenario horrified the council officers, who claimed that it was more appropriate for (paid) professionals to take up these roles. This illustrates the risk aversion typical of new initiatives (Taylor, 2000) while also highlighting power differentials: elites were allowed, encouraged even, to retain positions of power. And so public accountability is sharpened rather than public participation enhanced (Newman, 2001).

COMMUNITY APPRAISAL GROUPS

Meanwhile getting involved in less highly defined activities provides a contrasting image of participation. In The Village group where the activity was initiated by local residents, participation and involvement were less complex and much more fluid. An initial public meeting held in March 1999 identified health as a key concern for the community. The project grew organically:

'It started off as a small survey, but it just grew like Topsy' (Local paper, 22.12.00).

This project received practical assistance from The Community Initiative, the local health authority, the local council and later from the local rural development agency.

'This project was originally envisaged as a simple questionnaire to identify health needs of all [The Village] residents. It quickly grew! Firstly because several surrounding villages look to [The Village] for health promotion activities it expanded to include four adjacent villages. Eventually the project explored the health needs and well being of different age groups' (Analysis of questionnaire, December 2000).

It was initiated from within the community and the group was very clear about being in control of the activities with which it got involved:

'I had a bit of a heated debate with the chair of the group today. We were going through the contents of the questionnaire over a cup of coffee at [Jim's] kitchen table. There was a clear absence of detailed housing questions. I pointed this out to [Jim] and suggested that they might like to include some specific questions that would constitute a housing needs survey [and so prevent them from having to conduct a separate survey at some later date if the need arose]. But [Jim] was fairly dogmatic about this. According to him this was their survey and they would include questions that they believed to be important, not just produce what he called a 'copycat' appraisal' (Research journal, 03.09.99).

Communities are constantly placed under pressure to conform to accepted wisdom about the rural development process. For those communities opting out of structured systems and choosing less orthodox approaches to development by operating in their own space and developing their own independent voice, options are reduced and in some cases can be seriously jeopardised (Jones, 2003). As Taylor points out, 'community players in these new spaces [need] to be sophisticated about their engagement and understanding of power' (2007:311). This was possible in The Village because a key figure was a retired community development professional. In the end housing emerged as a major issue in The Village to the extent that a housing sub-group was established. It had to conduct a separate housing needs survey to demonstrate demand so that the government planning department would allocate planning permission. In other words eventually it had to conform to social structures and regulations.

'Concrete evidence of actual housing need in [The Village] is now needed. Some will be obtained from residents through a questionnaire' (The Village group minutes, 02.10.01).

However the group was determined to hold few obligations, to retain control and to set the agenda:

'Instead of waiting for others to decide what local people needed, the aim was to ask local people themselves' (Local paper 22.12.00) and 'local ownership of the project was jealously guarded throughout' (Jim's paper on local strategic partnerships, 22.08.01).

Jim had a fundamental problem with the partnership approach where representatives of different agencies come together 'to devise strategies. These people are usually too distant in terms of wealth, culture, attitude and values from the people in greatest need to have any real idea of what would benefit them most' (Jim's paper on local strategic partnerships, 22.12.01).

A close association between the availability of funding or other support and the choice of technique can be problematic as it can result in a community undertaking a task simply because it is advocated by an external agency. The application of a particular technique within Great Villham was advocated by a professional even though sufficient resources did not exist for successful execution of that method. The process necessitated a lot of volunteer time and expertise and soon the technique itself became the focus of the group. Meetings focused on the problems associated with the process such as the lack of volunteers. Eventually those community members who were involved walked away from the process as they became frustrated with the seeming lack of progress and failure to deliver any tangible community projects. The questionnaires ended up in the offices of a rural development agency without a single shred of analysis (Professional practitioner, 04.04.02). Meanwhile the community group had long since disbanded but the CAP project was able to report to the County Community Issues Group that Great Villham had completed its appraisal (Meeting minutes, 09.10.00). The question of whether it was actually in the best interests of that community became a side issue and so a situation emerges where 'people deal with programme requirements at the expense of real issues' (Community champion, 24.04.02).

The Village Project group believed that while prioritising action within a broader programme of consultation was the best approach, the specific community appraisal methodology promoted across the county was inappropriate for them. It resisted applying a direct replica of the methodology and the members chose instead to write their own questions, rejecting the option of selecting questions from a template. It chose to operate from outside of the invited space, and by defining its own space occupied a more popular space (Cornwall, 2004), which was at one level in opposition to state led schemes. This was a deliberately political stance as a group member reflected: 'We didn't want to owe anybody anything' (24.04.02).

Consequently the support of the Community Action Plan team from the rural development agency was not available and this made the possibility of accessing funding from central government tricky as they recommended that communities use this approach. In due course when the group decided to apply for funding from government, a number of its members ensured that the application was worded in such a way that it was clear that it was conducting a community-based initiative that would be recognised in the same way as a Community Action Plan. The group was repackaged as one that fulfilled the requirements of those operating in the official regeneration domain. It suggested broad-based participation encompassing all interests within the community so that the survey covered 'social, economic, environmental and health interests . . . We want to draw up a Parish Plan covering housing, leisure and recreation, village design statements . . .' (application to the Countryside Agency, 13.02.02). Clearly by using the terms that would be understood by policymakers and government officials, the group had decided that at one level it was conforming with the system and choosing to participate in the sphere of rural regeneration. The language used in the application guidance was reflected back in the text of the application and the group subsequently received funding. However the decision to use an alternative, non-mainstream approach could have had devastating consequences for the project had the group been unwilling to engage with any of the regeneration rules and re-package their project in appropriate language. The rules of engagement exert powerful influences on the activity of the group, and less astute (Ward and McNicholas, 1998) or unsuitable groups (Herbert-Cheshire and Higgins, 2004) lose out. In the latter case they pursue the 'wrong' development strategies or respond through inappropriate means such as protest (Herbert-Cheshire and Higgins, 2004:300) and so are not eligible to access funding.

The origins of the Growthville appraisal group were less obvious, although they did come from within the community. Jack 'read somewhere about a village appraisal and sent for the books from Cheltenham and Gloucester. I asked [the council] and [the rural development agency] to get involved and then we had a public meeting . . . I then got agreement from the parish council to proceed and so I set up a separate group with responsibility for overseeing an appraisal' (Community champion, 01.05.02). However this was not the full story. A strategic partnership had been operating in the community for a number of years before this. Its original remit was to co-ordinate transport projects, but this expanded to include community projects.

> 'Another instalment from [Jack] re: some history to [Growthville]. Apparently the [Growthville] partnership has been around for a couple of years and primarily consists of reps drawn from local authorities and the PC. Transport is the key issue and with a budget of £100k a lot of improvement work is going on. It is hoped that findings from the appraisal will legitimise the work of the partnership. Whether or not this happens is quite another matter. [Jack] to send me more info.' (Research journal, 07.11.00).

It was this partnership that held funds received from the local council. They had been given money to spend on projects in the area, in agreement with the council. Were the appraisal group to have been an informal body operating without the support of the parish council it would have had severe problems accessing funds. The application form for the appraisal process specifically required that the submission was made in the town or parish council's name and also asked the applicant if the project had been discussed with the county-wide rural development agency. Even though the group was locally grown, it was involved with initiatives that were subject to rules that had been created outside of the community—in this case the conditions of the appraisal scheme and the regulations of the local council. Nonetheless councils often attempt to simplify the process wherever possible as the following case study illustrates.

Case Study: Spending the money

> *Unintentionally the system for appraising projects was designed with two opposing purposes. The Board wanted project appraisals to be as transparent, simple and accountable as possible. This was at odds with the actual SRB programme monitoring requirements that were complex and detailed. With these obligations in mind the Project Manager designed a four-page expression of interest and a fifteen-page full application form which was then approved by the Board. The former doubled as a full application to the Community Chest which provided funding of up to £2500 and was not subject to the rigorous appraisal and approval process of major projects (anything above £2500).*

> *'We [SRB Board] had a difficult dilemma over the process being pulled from a number of directions: being accountable, having*

systems in place whilst also getting on with the task in hand'
(30.11.00).

It could be argued that the ideals of transparency, accountability and
simplicity got lost in a mire of systems, procedures and regulations
as the appraisal system became complicated and protracted generat-
ing a high volume of paperwork. However the process reflected the
requirements of the RDA and entailed distinct appraisal and approval
stages.

The Board members held a powerful position as they made decisions
about how the Community Project funding should be allocated, albeit
within the SRB framework. Given the degree of expertise required to
participate as a full member of the Board, it may be unrealistic and
overambitious to expect that all individuals should move to this degree
of involvement, setting aside the issues of whether or not they wish to
get involved at such a level. The participation of individuals with little
experience of regeneration funding in the Board required the devel-
opment of particular skills to enable them to participate without the
existence of barriers.

Potentially many different projects within the geographic area could
have been funded through the scheme, with many different individu-
als benefiting. However as the discussion earlier points out, the ap-
plication process was not straightforward. Even for the Community
Chest funds (providing grants up to £2500) the applicant had to
be confident and competent enough to complete a four-page form
and to obtain match funding, normally an amount equal to that
requested from the Chest. Any projects seeking more money than
this had to also complete a fifteen-page form. The Board members
(including the Project Manager and the Strand Co-ordinators) rec-
ognised this barrier and made it very clear that they would meet
with any prospective applicant to discuss the application process
and provide guidance where necessary. The Project Manager was
the primary source of application assistance, with Strand Co-ordi-
nators providing such help to a lesser degree.

So in both unstructured and structured activity, regeneration groups are
obliged to comply with certain institutional requirements.

Even though the Growthville project was identified from within the community, in a manner similar to circumstances in The Village initiative, a single
person seemed to be the driving force. As individuals they appeared to be
motivated by different things, but ultimately both groups sought to improve
'health, social, housing and transport services' (briefing notes for The Village

questionnaire). The process was slightly different, with The Village aiming for high levels of participation so that the group could 'work out, with the community, what to do to overcome these limitations and to get our voice heard and taken notice of in the planning processes of statutory and voluntary organisations' (ibid). I got the impression that group members enjoyed the process of participating. This was evidenced by different straplines appended to briefings for planning meetings, for example 'It should be an exciting meeting' (invitation to meeting, 19.09.00). In and of itself this would seem a fairly benign statement. But it actually reflected the attitude and the camaraderie of those involved.

It is significant that the progress of 'the group' relies on the commitment of a core of individuals each of whom spends a lot of time attending meetings or completing tasks between meetings. In The Village core attendance at planning meetings dwindled from twenty to just over half that number (Research journal, The Village meeting minutes). As attendance declined, so the amount of time that these people committed increased from monthly meetings, to often weekly gatherings. Originally scheduled for week nights, these became Saturday morning events. The upshot is that a core of individuals under the guise of a 'community group' often becomes the bedrock of the rural development activity. Croft and Beresford (1992) note, this is in contrast with the rhetoric of community development, which is of large-scale and broad-based involvement.

That aside, participation levels in both areas were very high with a 90% response rate to the questionnaire (letters sent from The Village group to local politicians 29.01.01) and just under 70% completing questionnaires in Growthville (Growthville Appraisal report, December 2000:3). In Growthville there was much less emphasis placed on the appraisal process with the participatory exercise perceived as a means to an end, with little evidence of community ownership. One of the stated objectives of the appraisal was

> 'To prepare and publish a final appraisal report for all households in the village and for this report to be sent to all the relevant public authorities' (Growthville Appraisal report, December 2000:3).

Indeed 'the steering group has not attempted to make any recommendations for action. The Parish Council . . . will undertake that role. [It] will begin the next phase of the initiative—that of setting up action groups to oversee the development of a number of ideas' (Growthville Appraisal report, December 2000:3, 15).

The conception of the Commuterville Project Association came about following a series of public meetings that were held to decide how to commemorate the New Millennium. Two projects were identified and agreed: one was to refurbish the village hall and the other was to build a

bridge linking two parts of the village (Community champion, 26.01.00).
The community had a genuine opportunity to identify a project of their
choosing and then to turn the idea into a reality. The two groups leading these projects were perceived by some to be in competition with each
other, but in the end they both succeeded in raising funds to complete
their objectives.

CONCLUSIONS

The individuals involved in a rural development group and driving forward
a local agenda may be unaware that they have been subject to complex
power relations. For example their perceived needs and desires may have
been manipulated to the extent that they have always been dependent on
the state doing things 'unto' them and have been unaware of their own
capacity to bring about change. Support is as much about enlightenment as
it is about education. Empowerment cannot be assumed (Schofield, 2002)
and so people need to be allowed to explore alternate ways of working.
This is crucial if real interests are to be identified, power relations understood and the relationship between agents and the state unpacked.

Participation in rural development structures demands the establishment of complex systems and procedures in order to meet even basic
requirements. It was shown to be a complex affair. An individual wishing to participate in the regeneration of Great Villham had the option
of doing so under the national SRB programme. This was not one that
emerged from within the community. In the language of development
programmes, it was a top-down approach. On the surface this would
seem to correlate to Lukes' expanded understanding of power (2005:65).
It relates to agents' (primarily the local authority) bringing about considerable effects by both furthering their own interests while in this case
also affecting the interests of others (the individuals and groups that got
involved in the Community Project). It is not yet clear to what extent the
interests of others were affected, if at all. We will return to this later.
What is apparent is that participation was tightly defined and left little
room for alternative approaches, such as projects that might challenge
the way in which the state chooses to frame particular issues. It limited
options for identifying the preferences, needs and the interests of the community. There was little space for negotiation and renegotiation so that
rather than being a dynamic process, it was fairly fixed.

Meeting funders' requirements can seem far removed from community
objectives and so challenge even the most committed voluntary members.
The Community Project Board focused on delivering the requirements of the
formal Delivery Plan document as the contract between regional development agency and the local authority demanded. Meanwhile smaller regeneration initiatives driven from within the community tended to focus on their

specific objectives and pursue small pockets of funding in a more ad hoc way. There is a danger that the funding framework over-rides the original driver for rural development. As groups pursue funding they become exposed to 'rules' of rural development with the risk that their original sense of purpose is diluted or sidetracked. The rules of engagement inevitably influence, in some cases dictate, the culture within the group thereby affecting interaction and individual gain from group membership. They also have a direct bearing on the membership of groups and the representation achieved within them as particular expertise is sought and experts are co-opted. 'Mimetic isomorphism' (Di Maggio and Powell, 1983:150) can result and so council norms and values, typically formal and prescribed, are replicated (see also Chapter 8 for further discussion). This is at the expense of a group that is more casual and innovative. The latter structure is more attractive and enjoyable to community members who give up scarce time to invest in the group's activities. It is also more radical in terms of challenging the norms of the state as it is more likely to identify new solutions for ongoing problems, rather than replicating risk adverse projects.

In the existing framework it is difficult for groups to undertake alternative approaches. Even the smaller scale, bottom-up projects had to eventually comply with rural development structures in order to access funds for the implementation of their projects. The way in which they understood their problems was shaped by power relations. As The Village Project revealed despite strong attempts to forge ahead according to the local community's plan, when the time came to access funding, it had to repackage itself and to comply with the technical requirements of rural development structures. Nonetheless this type of bottom-up regeneration was a much more fluid affair than the top-down SRB initiative. There were more opportunities to participate and, ostensibly at least, to influence the agenda through questionnaires, public meetings and working groups.

It could be argued that participation is less of a problem than Arnstein's model would suggest. Individuals may not wish to participate at the heart of rural development structures; besides our analysis thus far suggests that this can be an onerous and complicated task requiring particular skills and expertise. Even though community-led projects achieved high levels of participation in terms of responses to the questionnaires, the initiatives appear to be led by individuals or community champions. The notion of participation is certainly confusing; perhaps it is something of a misnomer and it is less of a question of who is in and who is out (Wenger, 1999), and more one of who has power and who does not. Individuals derive their power from personal attributes such as skills, experience, social standing and employment status. Our initial enquiry suggests that these are underpinned within structures that comprise regeneration programmes. The state would appear to have a central role in defining these structures, so that it influences the way in which local groups come to frame their problems. It is not clear how much of a voice those on the periphery have within this process or the extent to which key individuals follow or set the agenda or indeed exert more subtle forms of power by shaping the regeneration framework. In other words the question still remains: what exactly are people participating in? These matters will be further investigated in the following chapter.

8 The Performance of Participation

The previous chapter showed how distinct participatory approaches can result in different possibilities for the rural development group. Participation was very clearly shown to be an ongoing dynamic process requiring constant attention. Repeated studies of community involvement have shown how communities have generally remained on the margins of power in partnership arrangements (Hastings, 1996; Taylor, 2000, Bochel, 2006; Gilchrist, 2006). In this chapter participatory practice is scrutinised to consider the reality of participation. It investigates where power is located in contemporary practices of regeneration. Just how much do communities, and individuals therein, participate in structures of rural development? How discernable and significant are power relations among rural development actors?

Given the different driving forces behind participatory techniques, it is hardly surprising that the practice of participation is uneven, 'with the community engagement rhetoric over the years far outpacing the reality of partnerships on the ground' (Taylor, 2007:298). Different groups experience a different quality of participation, and the voices and views of some groups are given greater weight than the voices of others (Edwards et al., 2000; Shortall, 2004). Participatory processes can be manipulative and harmful to those supposedly empowered (Cooke and Kothari, 2001; Hickey and Mohan, 2004). Equally the purpose of participation is irregular. Many groups place stronger emphasis on social and community development than economic development. For these groups social development is an end in itself, and in their eyes their activities are successful because they were providing for some previously unmet need (LRDP, 1994; Bryden et al., 1996; Hayward et al., 2004). Using the lens of power (Chapter 4) and micro-politics (Chapters 5 and 6) and building on the concept of participation as introduced in the previous chapter, this chapter considers the practice of participating in rural development and regeneration.

THE MOTIVATION FOR PARTICIPATION

It is clear that different individuals have different beliefs about the nature of involvement and as a consequence many techniques are used to achieve participation. Participatory approaches are implemented through a range of different agendas and instruments within modern society. Many international aid agencies advocate a community development paradigm using initiatives such as social investment funds, group-based micro-credit programmes and community-based natural resource management schemes (Babajanian, 2005). In its guidance on preparing Community Strategies (Office of the Deputy Prime Minister [ODPM], 2000), UK central government recommends that partnerships offer various opportunities for participation using techniques that apply to their local circumstances. This notion of flexibility is echoed in the literature with suggestions of a 'spectrum' (Williams, 2003:532) and 'repertoire' (Lowndes et al., 2001b:449) of participation. The use of various participatory techniques is sensitive to individual group and regional partnership approaches and so provides a means of legitimising many different forms of participation. Hence 'different participation approaches may be more suited to the needs of particular types of organisations' (Lowndes et al., 2001a:209).

Participatory processes often occur in tandem, be they the result of statutory obligations, the outcome of a proactive community or the consequence of regeneration or development agency activity. They are encouraged through policy initiatives and, as the case study below reveals, there are many and varied approaches. The upshot from these copious participatory techniques is that individuals living in any given area may be faced with endless opportunities to participate in activities. In short, different participatory schemes are superimposed and juxtaposed within a community. They create a dense network of structures that influence and are influenced by each other.

Case Study: Choosing participatory techniques

Selecting an appropriate technique is a significant task. Various good practice guides exist for this purpose including the New Economics Foundation's guide entitled Participation Works! (Lewis et al., 1998). This document directly addresses New Labour's Clause IV which promises a new system of government where communities are involved as much as possible in taking decisions that affect their area (Lewis et al., 1998)—hence the emergence of the governance approach. Participation Works (Lewis et al.) is an A-Z of participatory techniques from Action Planning (1998:7) to Team Syntegrity (1998:45). It recommends that for all participatory approaches groups consider issues such as values, involving the less articulate, the use of experts and the availability of resources—both time and money, before selecting their technique(s). For each technique methods are offered along with advice based on a Case Study. For instance Participatory Appraisal offers brainstorms, diagrams, community mapping and timelines among others as suggested methods to achieve an appraisal. Suggestions for resources required and options for further support in terms of contacts, publications and training are then provided (Lewis et al., 1998:33–34). This is as close as a group is likely to come to learning about a technique short of direct experience.

DIFFERENT VALUES, SHARED RESULTS?

Furthering the interest of the 'community' is a challenging issue. Even if the community could ever be neatly defined, the multiplicity of different interests that exist dictate that certain individuals are subject to power as their interests are sidelined.

Divided opinions emerged among group members in Growthville. One faction viewed the appraisal as an end in itself; it would improve life for the residents of the village by allowing them to work together and develop co-operative relationships. As a group they would raise issues among other agencies and eventually it would trigger further activity. The other group members, led by Jack, perceived the appraisal as a means of justifying the need for a doctor's surgery.

> 'During one of [Jack's] rants about the lack of health facilities I caught [Joan's] eye. For a brief moment there was a flicker of understanding as she rolled her eyes skywards. I assumed she was bored by [Jack's] incessant ranting! Not everyone felt the same way as he did and I started to wonder just how much of 'the community' shared his point of view. But at least I felt as though something finally clicked in place with the 'ladies' of [Growthville]' (Research journal, 03.10.00).

Shortly after this meeting the attendance of the woman and her friend became more sporadic. The lack of strong connections within the group proved to be critical to its progress. Thus while at some level these women felt that the overall objectives of the group were relevant to them, it emerged through discussion with them that they did not feel that the health issue should have been central to the focus of the group. They clearly felt powerless to influence the group's direction; the process was actually something of a facade. As we discovered in earlier chapters, other matters influenced the Growthville project, namely the personal agenda of a pivotal member and his desire to develop health facilities in the village. He failed to promote the message that the project was about raising funds locally or in partnership with others in order to implement changes within the community.

> 'For [Jack] this project is all about getting the GP surgery, I'm not really sure why he is pretending to do a full blown appraisal. In any case he doesn't seem to get that this is only the starting point of a much bigger series of projects' (Research journal, 29.03.00).

Jack retained influence, stifled debate and pursued his own personal interest whilst claiming to represent the community. He did not perceive power as something that could be shared. As a result he presented the exercise as one in which individuals could have their say and contribute to the group's overall direction. In reality the group was being used to achieve a particular end result: improvement of primary care facilities.

Failure to convey a strong message about the appraisal process and failure to delegate and share tasks formed a barrier to fuller participation. A further barrier existed because an influential member of the group fundamentally misunderstood the nature of power relations. He clearly believed that if he ceded power to others within the group by delegating then he would no longer be controlling the agenda and so he would be in a weaker position. However following from Mann (1986), it is clear that the overall power of the group would have been enhanced as it would have had access to a greater network of information, resources and contacts. Consequently it would be in a stronger position to lobby and campaign for enhanced healthcare facilities and it is more likely that Jack would have ultimately achieved his overarching objective.

In a similar vein we noted how at an early Forum meeting attendees misunderstood the agenda, believing that they could raise issues which they believed were important in their community. This was not actually the case, as they were participating in the Single Regeneration Budget (SRB) consultation process (see Chapter 7). Even though their participation meant that they were part of the process, they were constrained from acting freely and so the perception of involvement was not necessarily synonymous with its practice (Croft and Beresford, 1992). This was also the case for professional members in positions of power, as my reflections of the meetings indicate (I was vice-chair of the Board):

'[Tina] and I had a chat about the style of the [residents—council liaison] meeting. We were in complete agreement that [Rick's] style is very formal and that it may not be the most helpful in terms of getting the best out of people. Neither of us believe the format is right for the purpose of the group but we both feel powerless to do anything about it' (Research journal, 16.07.01).

There were further differences of ideology around the value of the rural development process, as the Community Project Manager explained:

'[Kate] explained her concerns to me. She perceives the project approval team as being fairly heavy handed in terms of economic outputs and deliverables. They have little or no experience of community development and don't understand the value of 'process'. Meanwhile the project development team are taking a community development approach. There is a real danger that projects developed by the strands will be bounced back to them for development or 'tightening up' to use [Rick's] phrase, [Kate] is worried that we will lose potential projects and fund only projects with economic outputs' (Research journal, 01.10.01).

The SRB rules raise questions relating to the purpose of participatory exercises and on the matter of means versus end. Although the official documents indicate the value of process issues such as 'capacity building', 'empowerment' and 'community involvement' (DETR, 1999b: para 1.2.5 and Annex D and 1999a: para 5.1.1) so that 'up to 10% of the total approved grant for a successful SRB scheme can be spent on capacity building projects over the life-time of the scheme' (DETR, 199b: para 1.4.8), it is not clear whether participation was valued as a process or simply as a means to an end. This ambiguity existed within the partnership:

> 'I had always taken on my roles in the [Community Project] on the understanding that this was an interim measure to someone from the community taking on this position. Thus capacity building, empowerment and all of the things that the [Community Project] was supposed to achieve would have happened. [The Council] do not see it this way— any time I remind them of these intentions they quickly state that that will not happen for some time as my skills and knowledge of the project remain vital to it achieving the defined outputs' (Research journal, 13.06.01).

This is ironic as unlike many of these types of projects, where no resources are allocated to shore up the development of partnership bids (Taylor, 2000), the SRB programme appeared to attempt to close the gap between rhetoric and reality. It allocated funds to capacity building and also allowed for a year zero whereby community involvement would become one of the main priorities (DETR, 1999b). Even so, bids had to quantify expected outputs over the course of the project, such as the numbers of jobs created.

In terms of attracting money into the community, The Village group was not as successful as the Community Project. However, if we measure success in terms of a group's ability to develop skills, provide positive experience and to educate external agencies (reverse capacity building) then perhaps The Village project was more successful. It also achieved a remarkable response rate (90%). But much of this was down to the individual chairing the group and his experience of the development process.

He implicitly understood the importance of the process: 'Groups need to go through phases that require a lot of support. The support is crucial otherwise the group will not progress' (Community champion, 24.04.02).

As I exited the research field in 2002, it was clear that this group had generated a lot of interest and had made many changes within the local area. A community transport and a health project were being implemented with plans afoot for an affordable housing scheme, a village design statement and enhanced leisure facilities.

The matter of process versus product was very much alive within House. As the project co-ordinator I was convinced of the intrinsic value of the project but for the senior management team (SMT) the project was instrumental to building more houses:

> 'It's interesting to get [Chris'] take on things. Basically he views the work that I'm doing as PR and essential to the ongoing and long term work of the association. He summarised it as saying that 'networking is key'. His focus now seems to be very much on the housing that comes as a result of these so called PR activities' (Research journal, 21.02.02).

This explains why throughout the research I was placed under pressure to get involved with a greater number of communities:

> 'I really think we need to be working with more groups. It will lend weight to the work of the project. It will help us to justify the final report, we'll be taken more seriously' (House SMT, 20.03.00).

Although the motivation for undertaking the Community Initiative may not have placed as much value on the rural development process as on the ultimate objective of building houses, the fact remained that House devoted resources to this type of activity. Other rural development agencies were under pressure to perform and to sustain a high profile in the sector in the course of their project activities, such as the Community Action Plans. The upshot was a plethora of professionals representing a number of agencies with different agendas, all operating in the same territory. That is a direct consequence of governance and is examined later in this chapter.

It is unlikely that groups will ever agree on the absolute merits of process and outcome, and perhaps such an agreement is unnecessary. But it would appear that in order to be successful the group must have a shared vision and agreement on where responsibilities lie, thereby concurring on the relative merits of process and on the end result.

RESPONSIBILITIES, MISCONCEPTIONS AND PARTICIPATION REALITIES

The reality of participation is that projects need visibility; individuals want to see things happening:

'If the community can see one positive change they are more likely to get involved and so it is important that the achievements are promoted and publicised as widely as possible. People have high expectations about what can be done, who will do it and over what timescale it will happen' (Community champion, 01.05.02).

The Community Project was wrapped in a complex system of processes and procedures, drawing together many different organisations and promoting an ethos of community participation. Scepticism remained among those involved on the nature of the power differentials among partners and the purpose of the involvement. A resident from one of the council estates had the following comment to make of the Community Project:

'But we're not really involved in this project. It's one of your schemes isn't it? I mean, we're not making decisions about how the money is spent are we?' (20.03.00)

Even those who were involved more fully were dubious about their capacity to exert influence. They believed that ultimately the Council was in control of the key decisions and so they were powerless to make change anyway. There was a feeling of 'remoteness' from the project and a tenant liaison officer stated that

'it is not clear how it fits with the other activities, it has been pigeon-holed. Stronger links with the refurbishment scheme [i.e. the revamping of the council estate] would be useful' (25.02.02).

Meanwhile reflecting on the state of the Board and my limited power as a member, I had the following thoughts:

'Things seem to be in turmoil within the Board itself and my own enthusiasm for involvement has waned somewhat. Procedures seem unclear; we seem to be focusing down on minute detail at the expense of the 'bigger picture'. The meeting at the end of June which I could not attend ignored all of the issues that I had specifically asked [Rick] to raise. It's incredibly frustrating having a [a key member] who is so strong willed and with such definite ideas about how to do things' (Research journal, 26.06.01).

Confusion existed about what people were getting involved in and where responsibilities lay:

'There is a lot of apathy as residents feel they have heard a lot of this before—a fuss is created but nothing happens and so they don't get involved in anything again. Also it is not always clear where responsibility lies, I don't believe the pc should do everything, but other agencies should be taking the lead for certain projects' (Community champion, 01.05.02).

'I don't expect things to happen very quickly, I know about the timescales involved and that the council has to raise funds. This appraisal is a long term project and I think we'll see changes in a few years time' (Growthville resident and House tenant, 20.08.01).

Uncertainty around responsibilities has the potential to damage positive community activities:

'At the meeting we discussed the format of the public exhibition (again) ... [Conor's] bottom line is that the residents of [Growthville] will have to take a decision about whether they would like to pay more taxes to fund increased services and facilities. If this question is posed at the meeting I can safely predict that nothing will happen. I don't want to pay any more taxes and I'm sure that few people in [Growthville] would like their taxes increased. I tried to explain as reasonably as possible that at this stage in the process funding should not be the issue. The discussion should be about identifying priorities and needs before moving on to think about creative funding solutions, most of which I would predict will not be from increased taxes' (Research journal, 13.02.01).

Bewilderment regarding responsibilities was not confined to fluid community activities, but also existed within pre-defined structures:

'Once again the meeting of the [Community Project] seemed to be excruciating in terms of the amount of detail covered. [Rick] goes into what I believe to be a lot of unnecessary detail and even though it might appear that the meetings will not take a very long time, we find ourselves attending 2 ½ hour sessions. How can these be effective, especially when they are held on a monthly basis? [Rick] dictates the whole style [of the meeting] and the result is a very butch and self-important format. Is this how he perceives a business meeting? I don't see how a shy, single mother who wanted to get involved in her community could play a role. But maybe I am missing something and the place for such person to get involved is not at the 'management' meetings but somewhere else. Nonetheless my understanding of the [Community Project] is that it is by the community for the community, so why are people like me running the board meetings? How can I / we ensure the ownership of the project by local people?' (Research journal, 21.11.01).

The boundaries between different agencies within the community regeneration groups may have been blurred, but there was no fuzziness around the issue of control and accountability in the SRB initiative. It was clear that the council was against handing over control:

> 'I spoke to [Council Officer] today re: SRB bid. He would like [the Community Initiative] to continue with co-ordinating community/social strand (alongside [the rural development agency]). I am happy to do this as it gives me something to get my teeth into. I also suggested to [Council Officer] that the SRB management group should have a 'voluntary representative' as it is purely made up of paid staff. Interesting to hear his comments, he does not want a lot of controversy and he considers that this might make things 'tricky'. However he agrees that the process should not be fully orchestrated and is happy for me to raise it if I wish. Also of significance, [Council Officer] told me that the Council wants to have ultimate control, in other words the Board would only make recommendations' (Research journal, 23.08.00).

The plan was to bring others, including key community leaders, on board through alternative but more relevant channels thus allowing the needs of the community to be fed into the process, albeit indirectly. The danger with this approach according to Wood et al. (2001) is that views can be represented quite differently when filtered through representatives compared to when they are expressed directly. Participation for the council officers was clearly a limited affair. Even if it had been promoted more extensively there is no guarantee that involvement would have been any greater.

Practical issues can interfere with good intentions, impeding involvement.

> 'I am interested in getting involved in community activities, but I work shifts and that means I can't get involved' (House tenant, 20.08.01).

> 'I like being involved but I can't give any more time to this group, I have two young children at home' (Resident of The Village, 23.02.02)

Some individuals believe in the value of participation for the good of their 'community' but for them personally there is no benefit; that is, the costs and benefits fall differentially (Cleaver, 2001). This was evident among residents who had not been involved in community appraisal projects:

> 'I would like to help with projects that interest me personally, such as helping to improve facilities for younger teenagers' (Growthville House tenant, 20.08.01).

In fact there was a group looking at this very issue, but this resident had not got involved. It would seem that while people like the idea of being involved, the reality is something quite different (Lowndes et al., 2001b).

Involvement in rural development activities can promise so much, expectations are easily raised. When people get involved they can find that the reality of participation is not quite what they expected. Securing ongoing participation and representation can be hampered by a range of seemingly small issues and these have a bearing on the eventual rural development activity. By limiting representation eventually the image of the group is projected as one where the usual suspects are involved while others are discouraged from ever becoming involved. As this occurs certain group members may also withdraw their support because they feel that the group does not represent their interests.

PLAYING THE REGENERATION GAME: RULES, REGULATIONS AND CONFORMING

Regeneration policy rhetoric is thick with references to partnership, participation, capacity building and empowerment. A quick review of the governing documents, the Single Regeneration Budget Guidance Manual—Regional Development Agencies (DETR, 1999a) and the SRB Round 6 bidding guidance (DETR, 1999b) reveals a less than simplistic programme exposing contradictions between policy and practice. These official documents provide guidance on how projects should be delivered—through instruments such as specific outputs and contractual arrangements—while crucial detail about *how* to get people involved and how to address power struggles is omitted (DETR, 1999b). 'How to' guides on public participation are notoriously terse on points of power and culture (De Souza Briggs, 1997).

In his specific example of the SRB, Schofield (2002) suggests that adaptation is required from managers to align their working practices to the needs of the community to make involvement a reality. There was a danger that the 'rules of the game' would exclude community voices (Taylor, 2003) as the practices of regeneration reflected those of the elite.

Power Differentials

From the outset it is clear that applications to the SRB require extensive expertise and organisational capacity. Taking the case of the accountable body, while 'Consideration should be given to building the capacity of other partners, particularly those from the voluntary and community sectors, to enable them to [lead the partnership]' (DETR, 1999b: para 1.3.2), 'the RDA [Regional Development Agencies] can only enter into a funding agreement with a legal entity capable of meeting the liabilities that flow from the conditions of grant' (DETR, 1999a: para 8.1). While guidance material did not compel partnerships to be led by local councils, their predominant role was promoted (DETR, 1999a) supporting the notion that the rules of engagement in partnerships are controlled by the public sector

(Taylor, 1999) and so rest firmly with the elite (Bochel, 2006). As a consequence the culture of the council was likely to be mimicked by that of the partnership as it evolved into a more formal structure in order to comply with managing the project.

Indeed in its offer letter to the Council the RDA gave the following instructions: 'demonstrate in your Delivery Plan how your Partnership will deliver the scheme and how . . . your Partnership will ensure effective management, monitoring and control of the SRB grant' (Offer letter, 02.08.00).

In turn the Council issued a paper 'Proposed Administration and Delivery Structure' (03.08.00). Discussing this in a Forum meeting, it was made clear that 'ultimate decision making powers lie with [the] District Council as they are the financially accountable body, but they will be advised by the Management Board, who will in turn be informed by the three Strands and the Consultation Forum' (24.08.00).

Reminiscent of Katz's claim that policy elites successfully define ideal types of organisational structures (1975), the Forum was told that 'the process [for managing the money] is now being finalised with [the] District Council' (07.03.01).

Somerville considers that this moulding of citizens in their own image or 'responsibilisation' reinforces and reproduces elite power (2005:125). Such 'mimetic isomorphism' (DiMaggio and Powell, 1983:150) is yet another example of Lukes' third dimension of power (2005). Participants in programmes are not necessarily aware of this subtle form of manipulation, and even if they are, their ability to effect change is questionable.

There was some awareness of the danger of the culture of the council being imposed upon the partnership. One of its officers considered using the local authority's council chamber for Community Project related meetings. But the working group considered that people might conclude, rightly or wrongly, that the local authority wished to retain power and its attempt of involving the community was merely a token gesture. Even ignoring this strong argument against its use, many other reasons not to use it were identified. The council chamber with microphones, fixed seating and wooden panelling was described as 'intimidating' even to those who were involved in many different types of meetings (Research journal, 13.10.99). The setting was staid, formal, overawing and suggestive of a bygone era. It is difficult to imagine how trusting relations encompassing reciprocity, moral obligations or even commitment might be achieved in such physically imposing surroundings. In addition given the central role of the officers, the style of the meetings was likely to mirror that of local authority committees, thus projecting an image that was inappropriate for a community-based partnership.

Ultimately meetings were held in a number of locations including local schools, the volunteer centre, a sheltered housing scheme and a business centre. This was a success in that each venue attracted different attendees and resulted in a variety of individuals putting forward their viewpoints.

Had the partnership only met in a single venue, it is debatable if the same level of participation would have been achieved.

Nonetheless the supremacy of the public sector, and the power wielded by it, was evident by the way in which capacity building was addressed. In separate interviews two different officers expressed the opinion that training programmes for the community would substantially assist their capacity to participate in the bid, but that ultimately Board membership and Strand leadership were administrative tasks requiring the skills of professionals.

Meanwhile the SRB guidance provides a section on 'Community Capacity Building' (DETR, 1999b: para 1.4.7), there is no similar section for other partners such as those from the public sector. And in case there is any doubt on this, the meaning of capacity building is spelled out as 'activities to increase the capacity of local communities to contribute to regeneration and the strengthening of the social fabric, for example through training of staff and volunteers in community groups, through the strengthening of networks, forums or representative structures' (DETR, 1999a: Appendix 1, output 8F). It is notable that this contrasts with international development programmes where funding is provided to local governments in Chile, Honduras, Bolivia and Zambia to build their capacity to effectively develop local-level initiatives (Babajanian, 2005). Also notable are the similarities to the participatory programmes of the late 1960s as observed by Arnstein (1969) where she refers to the one-way flows of information from officials to citizens. By assuming that capacity building is only required by certain partners in the SRB programme, the evolving structures emphasize the superiority of public sector approaches. There is little encouragement of change to their institutional practices through involvement in the programme. This inherent assumption of organisational superiority is a symptom of the privileged ruling stratum; it is part of the ideology of the elite (Mills, 1956).

Institutional Politics

A hierarchy of power within the governing structures was evident with the regional development agency ultimately retaining power over the SRB project. This was further complicated because of regionalisation and the resulting creation of new structures of governance. The newly formed RDAs inherited a number of funding programmes such as Rural Priority Areas[1] and the SRB from a range of central government agencies. Consequently many of the new RDAs were adjusting to internal, organisational issues around the creation of new teams and functions, while simultaneously endeavouring to fulfil statutory requirements within a strict timescale. SRB Round 6 was accompanied by a degree of disarray with this new administrative arrangement.

> 'The second SRB Board meeting of the month was held in order to get things moving and reach [the RDA's] date for the delivery plan. There

is some tension relating to what people would like to get out of the bid. [The RDA] has been less than helpful in this process and it has been giving conflicting guidance. It appears that our bid, probably due to its relatively small amount, has been allocated to a novice who is not quite sure of her own organisation's guidelines, etc. This officer has been changing her mind on a regular basis on the contents of the Delivery plan. This, I fear, is symptomatic of the problems of [the RDA]—no-one is quite sure of what they are doing' (Research journal, 27.09.00).

'SRB board meeting tonight and much discussion over £30k which will essentially be lost if it is not spent this year. Wonderful. We thought we had cracked it by putting it into a general fund for next year and we thought this had been agreed by [the RDA], but it turns out [the RDA] want it allocated and spent . . . This doesn't seem like a system that is amenable to inexperienced groups who wish to access SRB and who also have limited capacity. The whole system is basically flawed' (Research journal, 30.11.00).

Developing Structures, Manipulating Rules?

The bottom-up community projects were much more flexible in structure and organisation. Some of them had received funding from government agencies for the execution of community appraisals, and this required conforming to a particular set of requirements. However the amount of funding was much less than the SRB Community Project and correspondingly the system of reporting was much less onerous than the SRB. And although The Village project was not following a formal procedure, it had created a structure for activities. This was not imposed from outside the community, but came from within the group. A series of themed meetings was held during which priorities and action points were identified, in much the same way as for the Growthville group.

The group received small amounts of funding from the parish council, but they did not feel that lack of resources was a major barrier to their progress. They were not motivated to make an application for large amounts of funding. In fact one member believed that getting involved in programmes was not necessarily a good thing. I was keen to provide a small budget to the group, but this proved difficult to get rid of:

'I want to make a contribution of a few hundred pounds to the project (The Village), this is proving difficult. Why will groups not just accept money . . . getting rid of my funding is proving to be very tricky indeed' (Research journal, 15.11.00).

Actually Jim very clearly wished to retain control. He appreciated the ideals of the partnership approach:

'In theory, the consequence of LSPs appears to be beneficial to all concerned; resulting in more efficient use of resources and better co-ordination of services . . . unless local people are fully engaged from the outset . . . there is a danger that LSPs . . . will become nothing more than a top-down structure bearing little significance to those at the local level' (Jim's paper on strategic partnerships, 22.08.01).

Jim understood the power dynamic in the participatory process; he recognised the importance of setting the agenda and of retaining control in order to achieve independence. By accepting money from my project, he obviously felt that impediments would be created as the group would be beholden to House and its position compromised. He appreciated that

'having too many hurdles is off putting for people and acts as a deterrent' (01.05.02).

The group eventually had to participate in more formal rural development structures in order to make progress. It reached a point where it wished to explore the issue of housing and village planning in more detail. A popular scheme at the time was the Vital Villages' programme, where groups were able to access funding from a national government agency that would enable them to undertake a Village Design Statement. This had the potential to be a very strategic document as it could feed into a county-wide five-year plan, thereby formalising many of the issues that were identified as being important to the local community and highlighting them to agencies with responsibilities for service delivery in the area. Along with the officer from a local rural development agency we encouraged the group to apply for funding from this initiative. In the process of writing the application form we emphasised the need to widen group membership and to ensure that participation was maximised by extending invitations to all known voluntary organisations in the area. In fact we suggested that the group's application would be unsuccessful were this omitted. We used our knowledge of the rural development structure to exert power over a decision maker to encourage him to delegate and to relinquish some of his power.

'I offered to colour print 400 posters for the open day. One of the steering group members agreed to distribute these to all households in the village (c. 350). And individual invitations will go to all of the voluntary groups in the village—finally. This was only agreed to because [professional practitioner] and I insisted that it be written into the CA Vital Village application form, so now they have to do it. Sometimes sneaky tactics are called for' (23.02.02).

While astute applicants to regeneration funding frame their application in the language of the funding agent (Ward and McNicolas, 1998), the ability to understand and to manipulate the rules of rural development is a significant skill:

> 'In general we under-estimated the total outputs in the hope that this would not only reduce the amount of paperwork needed, but so that we could focus on the process of working with the community without being driven by lots of outputs. The outputs that SRB tend to use are very tangible with much of the focus on economic development' (Research journal, 13.02.01).

In other words cognizant of the SRB scheme, we curtailed the measurable project outputs. This was to free up the partnership to focus on establishing relationships, creating norms and so place it in a position to successfully manage the project. This expert or professional knowledge was important to the group, but professional input is not always viewed favourably. Achieving a suitable balance between professional involvement and community participation can be tricky as the next section illustrates.

The type of activities that a group embarks upon is affected by the rules of regeneration as this determines the funding that is available. This in turn impacts on the type of individuals who are attracted to, and become centrally involved in, the rural development process. Reliance on individuals who are familiar with the rules of the regeneration 'game' can result in a group replicating the activities of another community. At the same time individuals with creative ideas may be sidelined as power is bestowed on 'safe hands', those with relevant experience and perceived know-how. Hence positions of status are entrenched as powerholders retain their position and those marginalised struggle to attain power. Eventually the image of the group is projected as one where the usual suspects are involved and others are discouraged from ever becoming involved. As this occurs existing participation from certain group members may also diminish because they feel that the group does not represent the interests of the community.

Professionalisation

It is evident that on the one hand expertise is essential to tap resources but on the other hand, the majority remains on the edge of the process (Storey, 1999) and potentially their interests are not represented. Criticism has been waged against inter-agency regeneration initiatives that are dominated by professionals, often with little accountability or transparency to the local community (Hall and Mawson, 1999). But if we consider the rural development structure then it becomes clear that in order to participate professional skills are often not an option, but a prerequisite.

The administrative duties of the accountable body were not insubstantial; they amounted to a huge, technical task. They reflected the fact that auditing seemed to have more importance than radical political motivations. For instance duties included receipt of and use of the final grant payment; establishing effective project appraisal and financial management systems; drawing up ongoing evaluation plans; establishing a project for post scheme evaluation; submission of quarterly reports and annual audited reports to the RDA (DETR, 1999a, 1999b). Partnerships were required to demonstrate relationships with other schemes such as regional and national strategies, showing how the 'SRB contribution will enhance, reinforce and add value to other initiatives and public spending programmes' (1999b: para 1.5.1). Many of the obligations are depicted using technical phrases such as 'deed of novation' (1999a: Annex 1D) 'key indicators of performance', 'milestones' 'quantifiable outputs', and 'exit strategy' (1999a: para 5.1). With an emphasis on performance and on structuring in the manner of government bodies, the participatory process is both depoliticized and mainstreamed; and so the act of regeneration is technical rather than political (Taylor, 2003; Hickey and Mohan, 2004). We have evidence of a process where the agents of the elite, that is, those creating regeneration programmes, devise a structure that maintains the power of the ruling elite, that is, the policymakers and the central state (Mills, 1956). Meanwhile the parameters of policy are established beyond the community (Jones, 2003; Taylor, 2003). This can create an atmosphere of fear and result in risk aversion. The accountable body in the Community Project seemed to be anxious that it would make a mistake and this caused it to implement an overly scrupulous project management structure:

> '[The Council] seem to be scared of actually taking a decision, certainly helpful ones . . . Proof of this was the fact that they actually questioned how their processes compared to other SRB bids. To get the response 'very rigorous' from myself and [Edward], rigorous to the point that I can't see the reason. I'm not really sure where the caution is coming from, but this project is fast becoming unwieldy' (Research journal, 14.12.00).

There was a real danger that individual legitimacy in the wider community was compromised as they blended into the partnership. Overall their capacity to act within the partnership was restricted. By limiting the involvement of the majority to the margins, the power elite retained control of the institutional hierarchy of regeneration.

Even so it is worth considering the role of professionals in the process. Chapters 3 and 6 reveal how individuals may be driven by professional targets and so they are under pressure to shoehorn communities into particular programmes even if that community would be more suited to doing something else. As a result of this the participatory exercise may not be entirely proficient simply due to lack of experience as suggested by the following account from a council officer:

> 'My background is in council led business programmes. Engaging with the community was quite novel for me and I went through a steep learning curve. I was glad to have the [Community Initiative] there to share consultation techniques' (Council Officer, 25.02.02).

But sometimes there were so many professionals on the scene that they were jostling for position. The resentment from a CAP officer has already been noted (see Chapters 5 and 6), but at other times it was a simple case of there being too many individuals on hand.

> 'About thirty people showed up on the day. There were quite a lot of 'professionals' with some tension emerging among this group. But in the end I thought the day worked quite well. I managed to meet with some of the community representatives most of who seemed fairly positive about the work that was ongoing. I think I have made the connection with [Stanley] and now hopefully I will be drawn into the process a little bit more. Although there does seem to be quite a complement of professionals who have a remit to work in the area so there may only be a limited amount for me to do' (Research journal, 20.01.01).

> 'I spoke to [Jenny] about events in [Market Town]. There has been a recruitment drive for a number of different issue-based groups whose role it will be to lead a process. Despite concerted efforts on my part to be included in the mailing lists I still have not been successful. Maybe there are enough agencies providing help in [Market Town] and so there is no role for [the Community Initiative]—that's absolutely fine. I do find it strange because community champions don't often say 'no' to offers of help' (11.05.01).

Within this support process, the role of the individual was important. But this was about providing more than expert advice: 'continuity of key contacts is very important . . . individuals matter' (Community champion, 24.04.02).

It has been noted that individuals get involved in rural partnerships because they feel that they cannot afford to be left out or because they feel threatened by the partnership and do not want it to develop policies that will challenge their own interests (Edwards et al. 2001). In a similar way new forms of governance compel organisations to have a presence in the accompanying structures; otherwise they will not have access to funding opportunities and to further their organisational objectives. The array of professionals and accompanying intentions that this produces makes for a crowded environment as these individuals have specific skills and ideals as well as professional targets to meet. The efficacy of this approach must be questioned.

DEGREES OF COMMUNITY PARTICIPATION

Achieving 'full' community participation or even representation is an ongoing challenge. The Community Project structure was designed to give everyone in the project area a meaningful opportunity to influence the process. During interviews with residents at a Family Day celebration that was held to launch the refurbished play area, many of them were not aware of the SRB project.

> 'The what project? . . . no, I don't think I've heard of that, what's it for anyway?. . . . No I'm not involved with that, it sounds like a council scheme. I just live here' (Great Villham resident, 30.08.01).

This is suggestive of 'cliques' or 'higher circles', a characteristic of power elites (Mills, 1959:11) and of illegitimate power where real or implied consent to the SRB project was not necessarily given by individuals living in the community. These members of the local community rationally opted out of participation; they had little desire to be involved at the centre, choosing instead to participate on the periphery.

In The Village although a response rate of 80% was achieved in the questionnaires, initially it was a challenge to get local people interested in this community project. Jim felt that people were loath to give up time due to the pressures of modern life, and those with free time tended to be retired individuals who sought a social network. This was evident in the public meetings:

> 'I did feel a little out of place, there were few, if any under 50 [years of age]. I wondered if I really have anything in common with them and in fact if they were really able to represent the interests of this community' (Research Journal, 17.12.01)

One group member retained control of the group's activities, setting the agenda and failing to delegate responsibility to anyone else. That stated, he did understand the concept of meaningful participation and recognised that

> 'really the question should have been posed as 'what are we going to do for the community' rather than 'what do you need?' thus changing the focus of events' (24.04.02).

Eventually a series of themed meetings was held and the housing meeting attracted a lot of interest with over forty people attending. Housing is a notoriously contentious issue in rural England, with problems for local people finding suitable, affordable accommodation (Hoggart and Henderson, 2005). Meanwhile there is often a perception that 'outsiders' buy up property in villages, inflating the price for the local community. These issues were felt by residents in The Village.

'[Jim] kindly filled me in on some of the background to the housing issue. A number of big 6 bedroomed houses were being built in the middle of [The Village] and were then being sold for £300,000 as 'affordable housing' (Research journal, 18.01.01).

'What a success. [The Village] Project Group organised an open/public meeting on housing to which about 40 people showed up. This was an excellent turn out . . . the meeting was very positive. [Joe] explicitly acknowledged the inter-linkages between all of the topics that the [project] are dealing with, i.e. housing, transport, health and quality of life. There were lots of comments from the floor—a few people [were] very against newcomers it would seem. But on closer investigation they have no problem with newcomers as long as they don't take houses from local people. I got the impression from the discussion that most people at the meeting would actively welcome houses for local people being built in the village. They even identified suitable sites' (Research journal, 22.03.01).

The group arranged follow-up meetings with local government planning officers, landowners and housing associations. The housing issue was among one of the most successful for The Village project. Many of the other themed meetings were less successful with attendance deteriorating:

'[The Village] had a public exhibition today. Unfortunately a steady dribble, rather than a steady stream was the order of the day. Whilst not terribly encouraging for the steering group, it is difficult to know what to do about trying to increase support and take action on the different themes' (Research journal, 05.12.00).

I had tried unsuccessfully to get Jim to think about increasing interest and to address the matter of representation. He did not think this was an issue.

'I spoke to [Jim] who was preparing a press release before setting his mind to planning the next meeting. He still carries a lot of the responsibility and when I discussed this with him last night he seemed to believe that there was always someone who led the way' (Research journal, 27.04.01).

This seemed to be the case for all of the community projects. Jack clearly led the group in Growthville, while in Market Town Stanley assumed the role.

'[Stanley] seemed fairly pessimistic about the meeting proceeding, stating that if there were less than 15 people interested it would be cancelled. To help with the planning he had circulated invitations with reply slips for people to return . . . [Stanley's] wife provided what can only be described as a lavish spread of food for both lunchtime and for morning coffee and afternoon tea, all of which she had home-cooked that morning. I wonder what motivates [Stanley] and his wife? They don't even live in [Market Town] and yet they have committed so much time and effort into trying to get things to happen, they are the lynchpin for the regeneration projects' (20.01.01).

Whereas the assumption within rural development is often of boundless involvement, maximum participation and equal power, it may be the case that regeneration structures are better used to enable a certain type of participation such as that of 'peripheral insiders' (Maloney et al., 1994) in a particular enterprise with questionable power. All the while the professionals and the community champions retain fundamental control of the activity.

CONCLUSIONS

Participation is an irregular, complex process; it cannot be considered in extremes of all-or-nothing. In the case of the top-down regeneration scheme it was bounded by a series of rules set by agencies that reflected the chain of accountability within the programme. In the case of SRB funding this included central government, RDA and the lead agency which was ultimately accountable for the specific project. In the other community activities, while less rigidly prescribed, participation occurred within a particular structure and involved a particular community of individuals, both professional and voluntary, with an interest in regenerating that area. So for example participation within the Community Project represented an enabling role, helping the lead agency achieve its function of delivering the SRB funding and so fulfilling policy objectives of central government. But it did not require full participation of all residents within the area, only a certain degree of representation via the project structures. This raises questions about how interests are represented. Meanwhile the grassroots initiatives sought the opinion of the maximum number of residents through community appraisals, but a core of individuals, typically community champions, led the group. The extent of empowerment as a result of the development process is not easily judged, even if this was ever an objective.

Failing to give the wider 'community' an opportunity to take up positions of influence in the structures of governance devalues the status of the regeneration partnership. It also ensures that many individuals from the community are rendered powerless in terms of achieving goals that further their greater good. They are unable to make a choice about wielding influence or effecting change. Even those in positions of apparent power within the Community Project partnership structure were not entirely free to act according to inherent values and beliefs—they had to comply with the obligations of the SRB programme as set out by central government. The consequence is that the institutional roles that participants are allowed and expected to play determine their ultimate being (Mills, 1959). In other words, those participating as representatives of the community were playing out roles as regeneration actors, and the structure was such that they were unlikely to move beyond this.

That a 'power elite' exists is evidenced by the way in which the regeneration structure reflects the norms, values and expertise of policymakers and professional practitioners, underlining the supremacy of the power elite. This was shown to exist throughout the SRB process—from the design of the initial community consultation to the complex requirements of project management. In this way the regeneration elite define the structure and in turn source their power from that structure, all of which serves to reinforce power differentials. Even the locally grown projects conformed to certain structures within rural governance and they relied on a particular selection of individuals: an elite of professionals and community champions.

This chapter set out to consider where power is relocated in contemporary practices of regeneration, and it aimed to analyse the impact of this for rural communities. It shows how social power is evident throughout processes of community regeneration. It is 'distributed by the various enduring structural relationships in society and exercised by individuals and groups based on their location in a given structure' (Isaac, 1987:28). Agents were constrained by the structure of the funding framework so that while they were able to freely express their ideas and beliefs, to remain involved and to participate they had to abide by the complex rules of the regeneration game.

Participating in the regeneration game is a technical rather than a political exercise; it is not an all encompassing activity. The highly prescribed nature of this process meant that active involvement was limited by the boundaries of the programme, in this case the SRB. Such emphasis on an invited space (Cornwall, 2004) risked engaging with the usual suspects in a pseudo-participatory process to demonstrate compliance with programme requirements. Those outside the elite circle were thus confined to the margins and so their particular agendas were sidelined, as administrative concerns or projects that were deemed feasible by the regulations took precedence. The regeneration arena thus becomes inert and impotent (Williams, 2005) and ultimately potential achievements are curtailed.

This research suggests that the concept of community participation is problematic. It suggests that complex forms of power are being used to effect change via regeneration schemes. Whereas the assumption is of boundless involvement, maximum participation and equal power, it may be the case that regeneration structures are used to enable a certain type of participation depending on the degree of 'insiderness' and relating to degrees of access and influence (Maloney et al., 1994:26). To paraphrase Lukes (2005), the powerful [public sector] agents derive their capacity and legitimacy from structures to call on the obedience of less powerful [community and voluntary sector] agents. Public participation is not what it seems—it is a limited activity. It is limited in terms of what it is able to do and it is also limited in relation to the individuals who choose to get involved. Further, and crucially, the practice of regeneration exhibits many of the symptoms of an elite activity (Mills, 1959), being determined and controlled by a clique of policymakers and their corresponding agents, resulting in a process that does not necessarily represent the wider views of the community that it purports to represent. But as agencies in the 21st century increasingly channel resources through multi-level structures of governance, communities cannot afford to be left out.

On the one hand in this age of public accountability and scarcity of resources, it might be completely unrealistic and idealistic to consider that a community can have responsibility for entirely setting its own agenda, free from constraints, to determine endlessly the types of activities in which it can engage. It could also be argued that this is a benign form of power where coercion and violence were absent. On the other hand, it is entirely misleading to set up these structures and systems of governance and claim that they are acting wholly in the interest of the community; it is clear that while some interests are represented a great number lie outside of the development process. More modest assertions about the nature of regeneration practice are necessary and may lead to less scepticism. Crucially this is necessary if inequalities between localities and communities are to be overcome.

Maybe it would be over-simplistic to propose that government is tricking all partners in regeneration in a game of empowerment and participation, when really it is seeking to enhance accountability of the public sector right through to its citizens. But it may not be too far removed from the reality of the situation to counter the idea that community regeneration needs to redefine its parameters. Current models over-state their potential. They can never represent the needs of everyone in a community. Moreover, they tend to work with a particular group of individuals, an elite, who purport to represent broader interests and to achieve great things. The reality is that these regeneration partnerships operate in a controlled environment. Most individuals do not have full freedom to make choices about the nature of the activities with which they engage. They operate under the power of the government in the form of the policy elite who set the parameters of regeneration programmes which is then interpreted and acted out by agents of that elite. Consequently the regeneration landscape, rather than comprising of community regeneration, involves an elite regeneration community.

9   Conclusions

Rural development continues to be employed as a tool for community participation and involvement. Although the objectives may have shifted from radical community development of the 1960s that sought to steal power from the state, current rural development programmes nonetheless seek to empower citizens. Rural development policy today expects that local communities are in a position to identify solutions to challenges arising in their area and to participate in activities that address these problems. In this respect they have agency and a degree of autonomy. As a result we see the increasing popularity of models of governance across the globe.

This book illustrates the difficulties of implementing rural policy ideals. Using the lens of power it challenges the notion of community regeneration. The central argument made throughout this book is that community regeneration is a restricted activity consisting of a power elite.

Specifically it shows how rural development is a technical and involved process that typically relies on a few dedicated professional and voluntary activists. Perhaps there is little new in this assertion. But participation is an erratic, often lumpy affair. It differs between locales and between and within programmes. Activity can be initiated from within a community, or the impetus may be external. This research demonstrates how the rural development framework can encourage a type of participation that automatically excludes the masses. The framework or structure is defined by a policy elite in a manner that reflects the values, norms and working practices of that group. The customs and social practices of the non-elite are not always recognised. And so in order to participate, individuals or groups must comply with the terms of engagement, and respect the boundaries to action. If they are not equipped with the necessary skills or prepared to play by the rules of the game, their potential for involvement in these regeneration domains is limited. But perhaps more critically their potential to access public funds is jeopardised. So involvement itself is exclusive.

Individuals may therefore be excluded because they are incapable of participating to the levels that are necessary. However, they may also opt out from a position of choice; they may feel that quite simply participation is not for them or that they do not want to become involved in the structures of governance. The state and its policymakers have a responsibility to recognise that not all citizens wish to become empowered in this way. Individuals can derive power and satisfaction from other activities such as family structures, social networks or employment status. Claims about the potential impact of rural development must therefore be curtailed.

Within the bounds of regeneration the extent to which the power of the state has been altered is questionable. This book demonstrates how structures are inadequate to allow agents to ascertain their real interests at all times. Nevertheless within the boundaries of regeneration, those who are active are not always able to identify preferences and wants and so power is exerted. And so even for those involved in this elite activity, that participation may not be what it seems. The potential of collective action is constrained and the role of the state remains central. In the end central government continues to devise sectoral policies that may or may not complement territorial regeneration and development programmes. Further, the amount of funding that is channelled through these types of structures must be kept in perspective: it is often a relatively small portion of overall budgets. Thus the power of rural development projects and agents to achieve real and meaningful change is debatable.

Not wishing to dwell exclusively on negative features, the positive aspects of rural development must be underlined. Limited and exclusive rural development activity can bring many benefits to a given territory. However the restricted nature of this must be acknowledged by policymakers. These programmes do not represent a panacea for all rural problems. The state must retain a key position in supporting a wider policy framework that does not undermine locally based development. Sectoral policies need to complement rural development strategies. But more than this, the regeneration elite must consider more closely the needs of the masses. Otherwise, to restate Mills and Lukes, we are in danger of seeing rural development that is created in the likeness of the elite which ultimately is unable to properly meet the real interests of the masses.

Further, rural development has an intrinsic value. The positive value of participating in rural development was evident within the case studies. The importance of this cannot be underestimated. It is true that agents operate within a particular structure that determines their behaviour and their capacity for action. However, individuals also have a degree of autonomy. They are able to articulate ideas, demonstrate preferences and exert power. By developing the concept and importance of micro-politics and by considering Lukes' notion of power-with-a-face, this book has underlined the role of the individual within rural development. It shows how individuals use structures and personal attributes to exert power. Successful rural development processes rely on positive micro-political processes, but ultimately on social interaction between individuals each of whom has a particular set of traits, values and characteristics.

What does this mean for contemporary rural development? Questions must be asked about the exclusive nature of regeneration. Do nation-states really seek to use rural governance initiatives to shift power from the centre to the new partners of governance? Given the proportion of funding allocated to them, to what extent are rural development programmes able to address challenges within particular locales? Are regeneration programmes merely a sop to local elites? Do they supersede more radical community action? These issues will doubtless continue to arise in future debates on rural development. However it is clear that while policy increasingly relies on the apparatus of governance, limitations will prevail: risks will be avoided, opportunities will be lost and innovation curtailed. As a result some (elite) communities will fare better than others as inequalities persist.

Notes

NOTES TO CHAPTER 2

1. Formerly known as Rural Development Areas and managed by the Rural Development Commission, Rural Priority Areas operated using a partnership representing different stakeholders and providing funding for projects according to a strategic action plan and, as such, represented an integrated approach to rural development.
2. The RPA and SRB programmes were later terminated and funding streamlined to become a 'Single Pot' (DETR, 2001).
3. For example in East Anglia, staff from the Rural Development Commission were re-located to either the newly formed Countryside Agency or the East of England Development Agency. So in many cases the same faces were representing two very different organisations each of which was striving for a new identity.

NOTES TO CHAPTER 3

1. Department of Transport, Local Government and the Regions (DTLR) took over regeneration funding in 2000; this responsibility was then transferred to Department of Trade and Industry (DTI) following the June 2001 election (http://www.dti.gov.uk, last accessed 31.07.04.
2. In direct contrast to the way it is used here—that is, as the study of social life in real, naturally occurring settings—the term *naturalism* is also used by some social scientists (e.g. Bhaskar, 1989) to refer to the adoption of natural science models of research to the social sciences. This position is also known as positivism.
3. This is used in a different way to Atkinson (1990) who uses the same term to describe the creative rhetorical abilities of ethnographic writers.
4. The term *professionals* is used here to signify individuals who are engaged in an activity as their profession, that is, they earn an income from it, and frequently have received some level of training to enable them to practice. This is in contrast to the voluntary individual who practices regeneration in a voluntary capacity, but not in receipt of an income from those activities.

NOTES TO CHAPTER 4

1. It is known that this is not always so and examples of negative consequences have been highlighted (see for example Shortall, 2004; McAreavey, 2006).

NOTES TO CHAPTER 5

1. It should be noted that Fukuyama (1995) does recognise that societies with low levels of trust are also able to develop large corporations via government support.
2. This is unlike other interpretations of trust (see for example Fox, 1974; Seligman, 1992. Although Fox recognises the personal nature of trust, he conceives of institutionalised trust as rules (formal and informal understandings) and relations (communication, interdependence, authority). In this way he focuses solely on structured and institutionalised relationships.
3. Portes and Sensenbrenner (1993) and Portes (1998) provide a detailed analysis of social capital's relation to three tenets from classical sociological theory, describing links to 'value introjection' encompassing notions of social integration and of formal and substantive rationality after Durkheim and Weber respectively; to exchange and reciprocity after Simmel; and finally to bounded solidarity after Marx.

NOTES TO CHAPTER 8

1. Formerly known as Rural Development Areas and managed by the Rural Development Commission.

References

Adams, D. and Hess, M. (2001) Community public policy: Fad or fountain? Australian Journal of Public Administration Vol. 60 No. 2 pp. 13–23.

Adler, P.A. and Adler, P. (1999) The Ethnographer's ball—revisited. Journal of Contemporary Ethnography Vol. 28 No. 5 pp. 442–450.

Anderson, C.D. and Bell, M.M. (2003) The devil of social capital: a dilemma of American rural sociology. In Cloke, P. (ed) Country Visions, Essex: Pearson Education, pp. 232–234.

Anheier, H.K., Glasius, M. and Kaldor, M. (2001) Introducing global civil society. In Anheier, H.K., Glasius, M. and Kaldor, M. (eds) Global Civil Society, Oxford: Oxford University Press, pp. 3–22.

Arnstein S. (1969) A ladder of citizen participation. Journal of the American Institute of Planners Vol. 35 No. 4 pp. 216–224.

Atkinson, R. (2003) Addressing urban social exclusion through community involvement in urban regeneration. In Imrie, R. and Raco, M. (eds) Urban Renaissance? New Labour, Community and Urban Policy, Bristol: The Policy Press, pp. 109–119.

Babajanian, B.V. (2005) Promoting community development in post-Soviet Armenia: The social fund model. Social Policy and Administration Vol. 39 No. 4 pp. 448–462.

Bachrach, P. and Baratz, M.S. (1962) The two faces of power. American Political Science Review Vol. 56 pp. 947–952.

Barnes, B. (1988) The Nature of Power. Cambridge: Polity Press in association with Oxford: Basil Blackwell.

Barnes, M., Newman, J., Knops, A. and Sullivan, H. (2003) Constituting 'the public' in public participation. Public Administration Vol. 81 No. 2 pp. 379–399.

Baron, S., Field, J., and Shuller, T. (eds) (2000) Social capital: critical perspectives. Oxford: Oxford University Press.

Beattie, A. and Williams, F. Doha trade talks collapse, Financial Times 29.07.08 http://www.ft.com/cms/s/0/0638a320-5d8a-11dd-8129-000077b07658.html, last accessed 13.08.08.

Beck, U. (1999) What Is Globalization? Cambridge: Polity Press.

Bevir, M. and Rhodes, R.A.W. (2006) Defending interpretation. European Political Science vol. 5 pp. 69–83.

Bhaskar, R. (1989) The Possibility of Naturalism. Hemel Hemstead: Harvester.

Birt, C. (2007) A CAP on Health? The Impact of the EU Common Agricultural Policy on Public Health. London: Faculty of Public Health.

Blair, T. (1994) Sharing responsibility for crime. In Coote, A. (ed.) Families, Children and Crime, London: IPPR, p. 90.

Bloomfield, D., Collins, K., Fry, C. and Munton, R. (2001) Deliberation and inclusion: vehicles for increasing trust in UK public governance. Environment and Planning C Vol. 19 No. 4 pp. 501–513.

Blumer, H. (1969) Symbolic Interactionism: Perspective and Method. Englewood Cliffs, NJ: Prentice Hall.

Bochel, C. (2006) New Labour, participation and the policy process. Public Policy and Administration Vol. 21 No. 4 pp. 10–22.

Body-Gendrot, S. and Gittell, M. (eds) (2003) Social Capital and Social Citizenship. New York: Lexington Books.

Borger, J. US biofuel subsidies under attack at food summit, The Guardian, http://www.guardian.co.uk/environment/2008/jun/03/biofuels.energy, last accessed 03.10.08

Bourdieu, P. (1986) The forms of capital. In Richardson, J.G. (ed) Handbook of Theory and Research for the Sociology of Education, Westport, CT: Greenwood Press, pp. 241–258.

Brewer, J.D. (1994) The ethnographic critique of ethnography: sectarianism in the RUC. Sociology Vol. 28 No. 1 pp. 231–244.

Brewer, J.D. (2000) Ethnography. Buckingham: Open University Press.

Bridger, J.C. and Luloff, A.E. (2001) Building the sustainable community: is social capital the answer? Sociological Inquiry Vol. 71 No. 4 pp. 458–472.

Bryden, J. (2005) The OECD's new rural policy paradigm: multi-functionality and beyond. Paper presented at annual QUCAN meeting, Inverness 21–23 March.

Bryden, J., Watson, R.D., Storey, C. and van Alpen, J. (1996) Community Involvement and Rural Policy. Edinburgh: The Scottish Office.

Buller, H. (2000) Re-creating rural territories: LEADER in France. Sociologia Ruralis Vol. 40 No. 2 pp. 190–200.

Buller, H. and Wright, S. (eds) (1990) Rural Development: Problems and Practices. Hants, England: Avebury.

Burgess, R. G. (1984). In the Field: An Introduction to Field Research. London: Routledge.

Callanan, M. (2005) Institutionalising participation and governance? New participative structures in local government in Ireland. Public Administration Vol. 83 No. 4 pp. 909–929.

Cheverett, T. (1999) A research agenda for partnerships. In Westholm, W.,

Moseley, M. and Stenlås, N. (eds) Local Partnerships and Rural Development in Europe: A Literature Review of Practice and Theory, Falun: Dalarna Research Institute, pp. 103–128.

Cleaver, F. (2001) Institutions, agency and the limitations of participatory approaches to development. In Cooke B. and Kothari U. (eds) Participation: The New Tyranny? London: Zed Books, pp. 36–55.

Coghlan, D. and Brannick, T. (2001) Doing Action Research in Your Own Organisation. London: Sage.

Coleman, J.S. (1988) Social capital in the creation of human capital. The American Journal of Sociology Vol. 94 Issue Supplement: Organisations and institutions: sociological and economic approaches to the analysis of social structures pp. S95–S120.

Coleman, J.S. (1990) Foundations of Social Theory. Cambridge, MA: Belknap Press.

Commins, P. (2004) Poverty and social exclusion in rural areas: characteristics, processes and research issues. Sociologica Ruralis Vol. 44 No. 1 pp. 60–75.

Commission of the European Communities (CEC) (1988) The future of rural society. Brussels: Commission (88) 601, Final/2.

Commission of the European Communities (CEC) (1998) Agenda 2000: commission proposals. Brussels: Commission (98) 158, Final.

Commission of the European Communities (CEC) (2005) COUNCIL REGULA-
TION (EC) No 1698/2005 on support for rural development by the European
Agricultural Fund for Rural Development (EAFRD) OJ L277 pp. 1–40.
Cook, I. and Crang, M. (1995) Doing Ethnographies. Norwich: University of East
Anglia.
Cooke, B. and Kothari, U. (2001). The Case for participation as tyranny. In Cooke,
B. and Kothari, U. (eds) Participation: The New Tyranny? London: Zed Books,
pp. 1–15.
Cooley, C.H. (1942) Social Organisation. New York: Dryden Press.
Cornwall, A. (2004) New democratic spaces? The politics and dynamics of institu-
tionalised participation. IDS Bulletin Vol. 35 No. 2 pp. 1–10.
The Countryside Agency (2004) State of the Countryside Report. The Countryside
Agency.
Croft, S. and Beresford, P. (1992) The politics of participation. Critical Social Pol-
icy Vol. 12 No. 35 pp. 20–44.
Dahl, R.A. (1961) Who Governs? Democracy and Power in an American City. New
Haven, CT: Yale University Press.
Dasgupta, P. (1988) Trust as a commodity. In Gambetta, D. (ed) Trust: Making
and Breaking Co-operative Relations, Oxford: Basil Blackwell, pp. 49–72.
Dean, M. (1999) Governmentality: Power and Rule in Modern Society. London:
Sage.
Delanty, G. (2003) Community. London: Routledge.
Denny, C. (2001) CAP in hand is better than cap in hand, The Guardian Spe-
cial report: the countryside in crisis http://www.guardian.co.uk/business/2001/
aug/13/politics.ruralaffairs, last accessed 03.10.08
Denzin, N.K. and Lincoln, Y.S. (2000) Introduction: The discipline and practice
of qualitative research. In Denzin, N.K. and Lincoln, Y.S (eds) Handbook of
Qualitative Research, London: Sage, pp. 1–28.
Department of Education for Northern Ireland (DENI) (2006) Schools for the
Future: funding, strategy, sharing. Report of the independent strategic review of
education. Belfast: DENI.
Department for Environment, Food and Rural Affairs (DEFRA) (2000a) England
Leader+ Programme 2000–2006. London: HMSO.
DEFRA (2000b) England Rural Development Programme 2000–2006. London:
HMSO.
DEFRA (2000c) Our Countryside, Our Future: A Fair Deal for Rural England.
DEFRA (2000d) Our Countryside, Our Future: A Fair Deal for Rural England—A
summary. London: HMSO.
Department of Environment, Transport and the Regions (DETR) (1997) SRB Challenge
Fund: a handbook of good practice in management systems. London: HMSO.
DETR (1999a) Single Regeneration Budget: Guidance Manual. London: HMSO.
DETR (1999b) Single Regeneration Budget: Round Six Bidding Guidance. London:
HMSO.
DETR (2001a) Local Strategic Partnerships—Government Guidance. London:
HMSO.
DETR (2001b) Single Regeneration Budget: Guidance to Regional Development
Agencies on approval of successor schemes to the Single Regeneration Budget in
2001–02. London: HMSO.
Deshler, D. and Sock, D. (1985) Community Development Participation: A Con-
cept Review of the International Literature. Sweden: International League for
Social Commitment in Adult Education.
de Souza Briggs, X. (1998) Doing democracy up-close: culture, power and commu-
nication in community building. Journal of Planning Education and Research
Vol. 18 pp. 1–13.

De Tocqueville, A. (1969) Democracy in America (Mayer, J.P., ed). New York: Harper Perennial.

Dhesi, A.S. (2000) Social capital and community development. Community Development Journal Vol. 35 No. 3 pp. 199–214.

Di Maggio, P. and Powell, W.W. (1983) The iron cage revisited: institutional isomorphism and collective rationality in organisational fields. American Sociological Review Vol. 48 pp. 147–160.

Dunn, J. (1993) Back to the Rough Ground: 'Phronesis' and 'Techne' in Modern Philosophy and in Aristotle, Notre Dame: University of Notre Dame Press.

Durkheim, E. (1984) The Division of Labor in Society. New York: The Free Press.

Edwards, B. (1998) Charting the discourse of community action: perspectives from practice in rural Wales. Journal of Rural Studies Vol. 14 No. 1 pp. 63–77.

Edwards, B. and Foley, M. (1997) Social capital and the political economy of our discontent. American Behavioral Scientist Vol. 40 No. 5 pp. 669–678.

Edwards, B., Goodwin, M., Pemberton, S. and Woods, M. (2000). Partnerships working in rural regeneration. Governance and Empowerment. New York: Joseph Rowntree Foundation and Policy Press.

Edwards, B., Goodwin, M., Pemberton, S. and Woods, M. (2001). Partnerships, power, and scale in rural governance. Environment and Planning C: Government and Policy Vol. 19 No. 2 pp. 289–310.

Elwell, F. (1996a) The Sociology of Max Weber, April 15, 2004, http://www.faculty.rsu.edu/~felwell/Theorists/Weber/Whome.htm, last accessed 15.04.06.

Elwell, F. (1996b) Verstehen: Max Weber's home page. Available online at http://www.faculty.rsu.edu/~felwell/Theorists/Weber/Whome.htm, last accessed 05.04.06.

England, K. V. L. (1994) Getting Personal: Reflexivity, Positionality and Feminist Research. Professional Geographer. Vol 46. No. 1 pp.80–89.

Estlund, D. (1997) Beyond fairness and deliberation: the epistemic dimension of democratic authority. In Bohman, J. and Rehg, W (eds) Deliberative Democracy: Essays on Reason and Politics, Cambridge, MA: MIT Press, pp. 173–204.

European Commission (2004) Proposal for Council Regulation on support for rural development by the European Agricultural Fund for Rural Development (EAFRD) COM(2004)490 final. Brussels: European Commission.

Falk, I. and Kilpatrick, S. (2000) What is social capital? A study of interaction in a rural community. Sociologia Ruralis Vol. 40 No. 1 pp. 87–110.

Fielding, N. (1993) Ethnography. In Gilbert, G.N. (ed) Researching Social Life, London: Sage, pp. 154–171.

Fine, B. (2001a) It ain't social and it ain't capital. Research and Progress Vol. 1 Part 1 pp. 11–15.

Fine, B. (2001b) Perspectives: Social Capital and Realm of the Intellect. Economic and Political Weekly 03.03.01.

Fine, B. (2001c) Social Capital versus Social Theory: Political Economy and Social Science at the Turn of the Millennium. London: Routledge.

Fine, B. (2003) Social capital: The World Bank's fungible friend [Review essay]. Journal of Agrarian Change Vol. 3 No. 4 pp. 586–603.

Flora, J.L. (1998) Social capital and communities of place. Rural Sociology Vol. 63 No. 4 pp. 181–506.

Flyvbjerg, B. (1998) Rationality and Power: Democracy in Practice. Chicago: University of Chicago Press.

Foley, M.W. and Edwards, B. (1997) Editors' introduction: Escape from politics? Social Theory and the Social Capital Debate. American Behavioural Scientist Vol. 40 No. 5 pp. 550–561.

Foucault, M. (1980) Power/Knowledge. Selected Interviews and Other Writings 1972–1977. New York: Pantheon Books.

Foucault, M. (1988) The Ethic of Care for the Self as a Practice of Freedom: an Interview with Michel Foucault on 20 January 1984. In Bernauer, J. and Rasmussen, D. (eds) The Final Foucault, Cambridge, MA: MIT Press. Trans. by Gauthier, J.D. pp. 1–20.

Foucault, M. (1990) The History of Sexuality Volume 1: An Introduction. Translated by Robert Hurley. New York: Vintage Books.

Foucault, M. (1991) Governmentality. In Burchell, G., Gordon, C. and Miller, P. (eds) The Foucault Effect: Studies in Governmentality, London: Harvester Wheatsheaf, pp. 87–104.

Fox, A. (1974) Beyond Contract: Work, Power and Trust Relations. London: Faber and Faber.

Fox, J. and Gershman, J. (2000) The World Bank and social capital: lessons from ten rural development projects in the Philippines and Mexico. Policy Sciences Vol. 33 Nos. 3–4 pp. 399–420.

Fukuyama, F. (1995) Trust: The Social Virtues and the Creation of Prosperity. New York: The Free Press.

Fukuyama, F. (2002) Social capital and development: the coming agenda. SAIS Review Vol. 22 No. 1 pp. 23–37.

Gambetta, D. (1988) Trust: Making and Breaking Co-operative Relations. Oxford: Basil Blackwell.

Gaventa, G. (2004) Towards participatory governance: assessing the transformative possibilities. In Hickey, S. and Mohan, G. (eds) Participation: From Tyranny to Transformation? London: Zed Books, pp. 25–42.

Giddens, A. (1968) 'Power' in the recent writings of Talcott Parsons. Sociology Vol. 2 pp. 257–272.

Giddens, A. (1979) Central Problems in Social Theory. Berkeley: University of California.

Giddens, A. (1984) The Constitution of Society. Cambridge: Polity Press.

Giddens, A. (1998) The Third Way: The Renewal of Social Democracy. Cambridge: Polity Press.

Gilchrist, A. (2006) Partnership and participation: power in process. Public Policy and Administration Vol. 21 No. 3 pp. 70–85.

Gillespie, G.J. and Sinclair, P.R. (2000) Shelves and bins: varieties of qualitative sociology in rural studies. Rural Sociology Vol. 65 No. 2 pp. 180–193.

Gittell, M., Newan, K. and Pierre-Louis, F. (2001) Empowerment Zones: An opportunity missed. A six-city comparative study. New York: The Howard Samuels State Management and Policy Centre.

Goffman, E. (1959) The Presentation of Self in Everyday Life. Garden City, NY: Doubleday.

Goodwin, M. (1998) The governance of rural areas: some emerging research issues and agendas. Journal of Rural Studies Vol. 14 No. 1 pp. 5–12.

Goodwin, M. (2003) Partnership working and rural governance: issues of community involvement and participation. Paper presented to Social Exclusion and Rural Governance Seminar—DEFRA, ESRC and Countryside Agency 28/02/03.

Goodwin, M. (2006) Multi-level governance in rural UK: recent debates. Paper presented to ESRC New Rural Economies Seminar Series 13/01/06.

Granovetter, M. (1985) Economic action, social structure and embeddedness. American Journal of Sociology Vol. 91 No. 3 pp. 148–510.

Grillo, R.D. (2001) Transnational migration and multiculturalism in Europe. Oxford: ESRC Transnational Communities Working Paper WPTC-01–08.

Grootaert, C., and van Bastelaer, T. (eds) (2002) with foreward by Putnam, R.D. The Role of Social Capital in Development. Cambridge, MA: Harvard University Press.

Hajer, M.J. and Wagenaar, H. (2003) Introduction. In Hajer, M.J. and Wagenaar, H. (eds) Deliberative Policy Analysis. Understanding Governance in the Network Society, pp. 1–30. Cambridge: Cambridge University Press.

Hall, S. (1980) Nicos Poulantzas: State, power, socialism. New Left Review, No. 119, pp. 60–69.

Hall, S. and Mawson, J. (1999) Challenge funding, contracts and area regeneration. Bristol: The Policy Press.

Hammersley, M. (1990) Reading Ethnographic Research. London: Longman.

Hammersley, M. amd Atkinson, P. (1995) Ethnography: Principles in practice (2nd ed.) London: Routledge.

Harriss, J. and de Renzio, P. (1997). Policy Arena: Missing link or analytically missing?: The concept of social capital. Edited by John Harriss. An introductory bibliographic essay. Journal of International Development Vol. 9, No. 7 pp. 919–937.

Harvey, D.R. (2004) Policy dependency and reform: economic gains versus political pains. Agricultural Economics Vol. 31 No. 2–3 pp. 265–275.

Hastings, A. (1996) Unravelling the process of 'partnership' in urban regeneration policy. Urban Studies Vol. 33, No. 2 pp. 253–268.

Hayward, C.R. (1998) De-facing power. Polity Vol. 23 No. 1 pp. 1–22.

Hayward, C.R. (2000) De-facing Power. Cambridge: Cambridge University Press.

Hayward, C., Simpson L. and Wood, L. (2004) Still left out in the cold: Problematising participatory research and development. Sociologia Ruralis Vol. 44 No. 1 pp. 95–108.

Heiskala, R. (2001) Theorising power: Weber, Parsons, Foucault and neostructuralism. Theory and Methods Vol. 40 No. 2 pp. 241–264.

Herbert, S. (2000) For Ethnography. Progress in Human Geography. Vol. 24 No. 4 pp. 550–568.

Herbert-Cheshire, L. and Higgins, V. (2004) From risky to responsible: expertise and the governing of community-led rural development. Journal of Rural Studies Vol. 20 No. 3 pp. 289–302.

Hewison, K. (2002) The World Bank and Thailand: Crisis and social safety nets. Public Administration and Policy Vol. 11 No. 1 pp. 1–22.

Hickey, S. and Mohan, G. (2004) Towards participation as transformation: critical themes and challenges. In Hickey, S. and Mohan, G. (eds) Participation: From Tyranny to Transformation, London: Zed Books, pp. 3–24.

Hindess, B. (1996) Discourses of Power: From Hobbes to Foucault. Oxford: Blackwell Publishers.

Hoggart, K. and Henderson, S. (2005) Excluding exceptions: Housing non-affordability and the oppression of environmental sustainability? Journal of Rural Studies Vol. 21 No. 2 pp. 181–119.

Hoonaard, W.C. van den (2003) Is anonymity an artifact in ethnographic research? Journal of Academic Ethics Vol. 1 No. 2 pp. 141–151.

Hunter, F. (1953) Community Power Structure. Chapel Hill: University of North Carolina.

Ingham, H., Ingham, M. and Weclawowicz, G. (1998) Agricultural reform in post-transition Poland. Tijdschrift voor Economische en Sociale Geografie Vol. 89 No. 2 pp. 150–160.

Institute of Public Policy Research (IPPR). (1998) Leading the way: a new vision for local government by the Rt. Hon. Tony Blair MP. IPPR.

Isaac, J.C. (1987) Beyond the three faces of power. Polity Vol. 20 No. 1 pp. 4–31.

Jessop, B. (1990) State Theory: Putting the Capitalist State in Its Place. Cambridge: Polity Press.

Jessop, B. (2002) The Future of the Capitalist State. Cambridge: Polity Press.

Jessop, B. (2005) The political economy of scale and European governance. Royal Dutch Geographical Society KNAG Vol. 96 No. 2 pp. 225–230.

Jones, O. and Little, J. (2000) Rural challenge(s): partnership and new rural governance. Journal of Rural Studies Vol. 16 No. 2 pp. 171–183.

Jones, P. (2003) Urban regeneration's poisoned chalice: is there an impasse in (community) participation-based policy? Urban Studies Vol. 40 No. 3 pp. 581–601.

Katseli, L.T., Lucas, R.E.B., and Xenogiani, T. (2006) Policies for migration and development: A European perspective. Policy Brief No. 30 Paris: OECD.

Katz, M.B. (1975) Class, Bureaucracy and Schools: The Illusion of Educational Change in America. New York: Berger.

Kawatchi, I., Kimberly, L. and Prothrow-Smith, D. (1997) Social capital, income inequality and mortality. American Journal of Public Health Vol. 87 No. 9 pp. 1491–1498.

Keane, J. (2003) Global Civil Society? Cambridge: Cambridge University Press.

Kotter, J. (1985) Power and Influence: Beyond Formal Authority. New York: The Free Press.

Krebs A. (1997) Discourse ethics and nature. Environmental Values Vol. 6 No. 3 pp. 269–279.

Latour, B. (1986) The powers of association. In Law, J. (ed) Power, Action and Belief: A New Sociology of Knowledge, London: Routledge, pp. 264–280.

Lee, J., Árnason, A., Nightingale, A. and Shucksmith, M. (2005) Networking: social capital and identities in European rural development. Sociologia Ruralis Vol. 45 No. 4 pp. 269–283.

Levitas, R. (1998) The Inclusive Society? Social Exclusion and New Labour. London: Macmillan.

Levitas, R. (2000) Viewpoint: Community, utopia and New Labour. Local Economy Vol. 15 No. 3 pp. 188–197.

Lewis, J., Walker, P. with Unsworth, C. (eds) (1998) Participation Works! 21 Techniques of Community Participation for the 21st Century. London: New Economics Foundation.

Locke, J. (1979) An Essay Concerning Human Understanding (Nidditch, P.H., ed). Oxford: Clarendon Press.

Lowe, P. and Brouwer, F. (2000) Agenda 2000: a wasted opportunity? In Brouwer, F. and Lowe, P. (eds) CAP Regimes and the European Community, Wallingford: CAB International, pp. 321–334.

Lowe P., Buller, H. and Ward, N. (2002) Setting the next agenda? British and French approaches to the second pillar of the Common Agricultural Policy. Journal of Rural Studies Vol. 18 No. 1 pp. 1–17.

Lowndes, V. (1999) Management change in local governance. In G. Stoker (ed) The New Management of British Local Governance, London: Macmillan, pp. 22–39.

Lowndes, V., Pratchett, L. and Stoker, G. (2001a) Trends in public participation: Part 1—Local government perspectives. Public Administration Vol. 79 No. 1 pp. 205–222.

Lowndes, V., Pratchett, L. and Stoker, G. (2001b) Trends in public participation: Part 2—Citizens' perspectives. Public Administration Vol. 79 No. 2 pp. 445–455.

Lowndes, V. and Skelcher, C. (1998) The Dynamics of multi-organisational partnerships: an analysis of changing modes of governance. Public Administration Vol. 76, No. 2 pp. 313–333.

Lowndes, V. and Wilson, D. (2001) Social capital and local governance: exploring the institutional design variable. Political Studies Vol 49 pp. 629–647.

Lowry, K., Adler, P. and Milner, N. (1997) Participating the public: group process, politics and planning. Journal of Planning Education and Research Vol. 16 No. 3 pp. 177–187.

Local and Regional Development Planning (1994) A review of the Northern Ireland Rural Development Council. Cookstown: RDC (unpublished).

Luhmann, N. (1980) Trust and Power. New York: Wiley.

Luhmann, N. (1988) Familiarity, confidence, trust: problems and alternatives. In Gambetta, D. (ed) Trust: Making and Breaking Cooperative Relations, Oxford: Blackwell, pp. 95–107.

Lukes, S. (2005) Power: A Radical View (2nd expanded edn). London: Palgrave Macmillan. (Original work published in 1974)

MacKinnon, D. (2002) Rural governance and local involvement: assessing state-community relations in the Scottish Highlands. Journal of Rural Studies Vol. 18 No. 3 pp. 307–324.

Maloney, W.A., Smith, G. and Stoker, G. (2000) Social capital and associational life. In Baron, S., Field, J. and Schuller, T. (eds) Social Capital: Critical Perspectives, Oxford: Oxford University Press, pp. 212–225.

Maloney, W.J., Jordan, G. and McLaughlin, A. (1994) Interest groups and public policy: the insider/outsider model revisited. Journal of Public Policy Vol. 14 No. 1 pp. 17–38.

Mann, M. (1986) The Sources of Social Power, Vol. 1: A History of Power from the Beginning to AD 1760. New York: Cambridge University Press.

Marinetto M. (2003) Who wants to be an active citizen? The politics and practice of community involvement. Sociology Vol. 31 No. 1 pp. 103–120.

Marsden, T. (2003). The Condition of Rural Sustainability. Assen: Royal Van Gorcum.

Marsden, T. and Sonnino, R. (2008) Rural development and the regional state: denying multifunctional agriculture in the UK. Journal of Rural Studies Vol. 24 No. 4 pp. 422–431.

Marx, K. (1959) Economic and Philosophical Manuscripts of 1844 Moscow: Progress Publishers, http://www.marxists.org/archive/marx/works/1844/manuscripts/preface.htm, last accessed 18.07.07.

Marx, K. (1964) Economic and Philosophic Manuscripts of 1844. New York: International Publishers.

Marx, K. and Engels, F. (1845) The German Ideology, http://www.marxists.org/archive/marx/works/1845/german-ideology/ch01.htm, last accessed 03.07.07

Mayer, M. (2003) The onward sweep of social capital: causes and consequences for understanding cities, communities and urban movements. International Journal of Urban and Regional Research Vol. 12 No. 1 pp. 110–132.

Mayerfield Bell, M. (1994) Childerley: nature and morality in a country village. Chicago and London: University of Chicago Press.

McAreavey, R. (2003) Sustaining Rural Communities. London: The Countryside Agency Publications.

McAreavey, R. (2006) Getting close to the action: the micro-politics of rural development. Sociologia Ruralis Vol. 46 No. 2 pp. 85–103.

McAreavey, R. (2007) A hidden cost? Rural development group politics. Euro-Choices Vol. 6 No. 1 pp. 38–43.

McAreavey, R. (2008) Researcher and employee: Reflections on reflective practice in rural development research. Sociologia Ruralis Vol. 48 No. 4 pp. 389–407.

McDowell, L. (1992) Valid games? A response to Erica Schoenberger. Professional Geographer Vol. 44 No. 2 pp. 212–215.

McGrath, B. (2001) 'A problem of resources': defining rural youth encounters in education, work and housing. Journal of Rural Studies Vol. 17 No. 4 pp. 481–495.

Mead, G.H. (1934) Mind, Self and Society from the Standpoint of Social Behaviourism. Chicago: University of Chicago Press.

Miliband, R. (1969) The State in Capitalist Society. London: Weidenfeld and Nicolson.

Mills, C.W. (1956) The Power Elite. New York: Oxford University Press.

Mills. C.W. (1959) The Sociological Imagination. Oxford: Oxford University Press.

Molyneaux, M. (2002) Gender and the silences of social capital: Lessons from Latin America. Development and Change Vol. 33 No. 2 pp. 167–188.

Montgomery, J. (2000) Social capital as a policy resource. Policy Sciences Vol. 33 Nos. 3–4 pp. 227–243.

Morris, J. (1981) Synthesis of Integrated Rural Development Projects, Evaluation Study 438, Overseas Development Administration, London.

Morriss, P. (2002) Power: A Philosophical Analysis (2nd edn). Manchester: Manchester University Press. (Original work published in 1987)

Mowbray, M. (2005) Community capacity building or state opportunism. Community Development Journal Vol. 40 No. 3 pp. 255–264.

Murdoch J. and Abram, S. (1998) Defining the limits of community governance. Journal of Rural Studies Vol. 14 No. 1 pp. 41–50.

Murdoch, J. and Ward, N. (1997) Governmentality and territoriality: the statistical manufacture of Britain's 'national farm'. Political Geography Vol. 16 No. 4 pp. 307–324.

Murtagh, B. (2001) Partnerships and area regeneration policy in Northern Ireland. Local Economy Vol. 16 No. 1 pp. 50–62.

Newby, H. (1977) The Deferential Worker. London: Allen Lane.

Newman, J. (2001) Modernising Governance: New Labour, Policy and Society. London: Sage.

Newton, K. (2001) Trust, social capital, civil society, and democracy. International Political Science Review Vol. 22 No. 2 pp. 201–214.

Northern Ireland Executive (2006) Better government for Northern Ireland. Final decisions of the Review of Public Administration. Belfast: NI Executive.

Northern Ireland Rural Development Programme 2007–2013 (2007) Belfast: Department of Agriculture and Rural Development.

Oakley, P. (1991) Projects with People. The Practice of Participation in Rural Development. International Labour Office (via Intermediate Technology Publishing, London).

Office of the Deputy Prime Minister (ODPM) (2000) Preparing Community Strategies: Government Guidance to Local Authorities. London: HMSO.

Organisation for Economic Co-operation and Development OECD (1999) Policy-making for predominantly rural regions: concepts and issues. Working Party on Territorial Policy in Rural Areas. Working Document DT/TDPC/RUR (99)2 Paris: OECD.

OECD (2001a) Local Partnerships for Better Governance. Paris: OECD.

OECD (2001b) Multifunctionality: Towards an Analytical Framework. Paris: OECD.

OECD (2005) Modernising Government. The Way Forward. Paris: OECD.

OECD (2006) The New Rural Paradigm. Policies and Governance. Paris: OECD.

OECD (2008) Public Management Reviews—Ireland: Towards an Integrated Public Service. Paris: OECD.

Park, R.E. (1952) Human Communities: The City and Human Ecology. Glencoe IL: The Free Press.

Parsons, T. (1960) The distribution of power in American society. In T. Parsons Structure and Process in Modern Societies, Glencoe, IL: The Free Press, pp. 199–225.

Pearce, G., Ayres, S. and Tricker, M. (2005) Decentralisation and devolution in the English regions: Assessing the implications for rural policy and delivery. Journal of Rural Studies Vol. 21 Vol. 2 pp. 197–212.

Penninx, L., Spencer, D. and Van Hear, N. (2008) Migration and Integration in Europe: The State of Research. Oxford: COMPAS, University of Oxford.

Performance and Innovation Unit (1999) Rural Economies. London: The Cabinet Office.

Petit, M. (2008) The CAP after fifty years: a never-ending reform process. Euro-Choices Vol. 7, No. 2 p. 44.

Pierre, J. (2000) Introduction: understanding governance. In Pierre, J. (ed) Debating Governance: Authority, Steering and Democracy, Oxford: Oxford University Press, pp. 1–12.

Pini, B. (2004) On being a nice country girl and an academic feminist: using reflexivity in rural social research. Journal of Rural Studies Vol. 20 No. 2 pp. 169–179.

Pollard, N., Latorre, M. and Sriskandarajah, D. (2008) Floodgates or Turnstiles? Post-EU Enlargement Migration Flows to (and from) the UK. London: IPPR.

Porter, S. (1995) Nursing's Relationship with Medicine. Alderstot: Avebury.

Porter, S. (1998) Social Theory and Nursing Practice. Basingstoke: Macmillan.

Portes, A. (1998) Social capital: its origins and applications in modern sociology. Annual Review of Sociology Vol. 24 pp. 1–24.

Portes, A. (2000) The two meanings of social capital. Sociological Forum Vol. 15 No. 1 pp. 1–12.

Portes, A. and Sensenbrenner, J. (1993) Embeddedness and immigration—Notes on the social determinants of economic action. American Journal of Sociology. Vol. 98 No. 6 pp.1320–1350.

Poulantzas, N. (1969) The problem of the capitalist state. New Left Review No. 58, pp. 67–78.

Purdue, D., Razzaque, K., Hambleton, R., Stewart, M. with Huxham, C. and Vangen, S. (2000) Community Leadership in Area Regeneration. Bristol: The Policy Press.

Putnam, R.D. (1993) The prosperous community. The American Prospect Vol. 4 No. 13. pp. 35–42.

Putnam, R.D. (1995) Bowling alone: America's declining social capital. The Journal of Democracy Vol. 6 No. 1 pp. 65–78.

Putnam, R.D. (2000) Bowling Alone: The Collapse and Revival of American community. New York: Simon and Schuster.

Putnam, R.D. (2004) Education, Diversity, Social Cohesion and "Social Capital" Meeting of OECD Education Ministers, Raising the quality of learning for all. Dublin 18–19th March. http://www.oecd.org/dataoecd/37/55/30671102.doc, last accessed 06.10.08.

Putnam, R. D. and Goss, A. (2002) Introduction. In Putnam, R. (ed) Democracies in Flux: The Evolution of Social Capital in Contemporary Society, Oxford: Oxford University Press, pp. 3–19.

Raco, M. and Imrie, R. (2000) Governmentality and rights and responsibilities in urban policy. Environment and Planning A Vol. 32 Vol. 12 pp. 2187–2204.

Ray, C. (1998) Territory, structures and interpretation. Two case studies of the European Union's LEADER I programme. Journal of Rural Studies Vol. 14 No. 1 79–87.

Ray, C. (1999a) Endogenous development in an era of reflexive modernity. Journal of Rural Studies Vol. 15 No. 3 pp. 257–267.

Ray, C. (1999b)The Reflexive Practitioner and the Policy Process. Centre for Rural Economy Working Paper Series. Working Paper 40. pp. 1–29. Newcastle: University of Newcastle.

Ray, C. (2000) The EU LEADER programme: rural development laboratory Sociologia Ruralis Vol. 40 No. 2 pp. 163–172.

Rhodes, J., Tyler, P., Stevens, S., Warnock, C. and Otero-Garcia, M. (2002) Lessons and evaluation evidence from ten Single Regeneration Budget case studies. Mid term report. Department of Land Economy, Cambridge for DTLR. London: HMSO.

Rhodes, R.A.W. (1997) Understanding Governance: Policy Networks, Governance, Reflexivity and Accountability. Buckingham: Open University Press.

Ron, A. (2008) Power: A pragmatist, deliberative (and radical) view. The Journal of Political Philosophy Vol. 16 No. 3 pp. 272–292.

Rose, N. (1996) The death of the social? Re-figuring the territory of government. Economy and Society Vol. 25 No. 3 pp. 327–356.

Said, E. (1986) Foucault and the imagination of power. In Hoy, D.C. (ed) Foucault: A Critical Reader, Oxford: Basil Blackwell. pp. 149–155.

Sayer, A. (2004) Seeking the geographies of power. Economy and Society Vol. 33 No. 2 pp. 255–270.

Schofield, B. (2002) Partners in power: Governing the self-sustaining community. Sociology Vol. 36 No. 3 pp. 663–683.

Schuller, T., Baron, S. and Field, J. (2000) Social capital: A review and critique. In Baron, S., Field, J. and Schuller, T. (eds) Social Capital. Critical Perspectives, Oxford: Oxford University Press, pp. 1–38.

Scott, J.C. (1990) Domination and the Arts of Resistance: Hidden Transcripts. New Haven, CT: Yale University Press.

Scott, J. (2001) Power. Cambridge: Polity Press.

Scott, M. (2002) Delivering integrated rural development: insights from Northern Ireland. European Planning Studies Vol. 10 No. 8 pp. 1013–1025.

Scott, M. (2004) Building institutional capacity in rural Northern Ireland: the role of partnership governance in the LEADER II programme. Journal of Rural Studies Vol. 20 No. 1 pp. 49–59.

Seligman, A.B. (1992) The Idea of Civil Society. New York: The Free Press.

Servon, L.J. (2003) Social capital, identity politics, and social change. In Body-Gendrot, S. and Gittell, M. (eds) Social capital and social citizenship, New York: Lexington Books, pp. 13–21.

Shortall, S. (1990) Farm wives and power—an empirical study of the power relationships affecting women on Irish farms. PhD Thesis, National University of Ireland (unpublished).

Shortall, S. (1994) The Irish rural development paradigm—an exploratory analysis. The Economic and Social Review Vol. 25 No. 2 pp. 233–261.

Shortall, S. (2004) Social or economic goals, civic inclusion or exclusion? An analysis of rural development theory and practise. Sociologia Ruralis Vol. 44 No. 1 pp. 109–123.

Shortall, S. (2008) Are rural development programmes socially inclusive? Social inclusion, civic engagement, participation, and social capital: exploring the differences. Journal of Rural Studies Vol. 24 No. 4 pp. 450–457.

Shortall, S. and Shucksmith, M. (2001) Rural development in practice: issues arising in Scotland and Northern Ireland. Community Development Journal Vol. 36 No. 2 pp. 122–133.

Shucksmith, M. (2000) Endogenous development, social capital and social inclusion: perspectives from LEADER in the UK. Sociologia Ruralis Vol. 40 No. 1 pp. 72–87.

Shucksmith, M. (2008) Disintegrated rural development? Neo-endogenous rural development in an uncertain world—a prospectus for comparative research. Paper presented at annual QUCAN meeting, Ithaca, 10–13 June.

Simmel, G. (1950) The Sociology of Georg Simmel (Wolff, K., trans and ed). Glencoe, IL: The Free Press.

Skocpol, T. (1996) Unraveling from above. American Prospect Vol. 25 pp. 20–25.

Snow, D. A., Morrill, C. and Anderson, L. (2003) Elaborating analytic ethnography. Linking fieldwork and theory. Ethnography Vol. 4 No. 2 pp. 181–200.

Somerville, P. (2005) Community governance and democracy. Policy and Politics Vol. 33 No. 1 pp. 117–144.

Spedding, A. (2003) Village Appraisals. Briefing 86. Warwickshire: The Arthur Rank Centre.

Stanley, G.K., Marsden, T.K. and Milbourne, P. (2005) Governance, rurality and nature: exploring emerging discourses of state forestry in Britain. Environment and Planning C: Government and Policy Vol. 23 No. 5 pp. 679–695.

Stevenson, M. and Keating, L. (2006) Rural policy in Scotland after devolution. Journal of Regional Studies Vol. 40 No. 3 pp. 397–408.

Stewart, A. (1995) Two conceptions of citizenship. The British Journal of Sociology Vol. 46 No. 1 pp. 63–78.

Stoker, G. (1998) Governance as theory: five propositions. International Social Science Journal No. 155 pp. 17–28.

Stoker, G. (2005) New localism, participation and networked community governance. Paper presented at the 6th Global Forum on Reinventing Government 24–27/05/05, Seoul, Korea.

Storey, D. (1999) Issues of integration, participation and empowerment in rural development: the case of LEADER in the Republic of Ireland. Journal of Rural Studies Vol. 15. No. 3 pp. 307–315.

Strathern, M. (1982) The village as idea: constructs of villageness in Elmdon, Essex. In Cohen, A.P. (ed), Belonging, identity and social organisation in British rural cultures, Manchester: Manchester University Press, pp. 247–277.

Svendsen, G.L.H. and Svendsen, G.T. (2000) Measuring social capital: the Danish co-operative dairy movement. Sociologia Ruralis Vol. 40 No. 1 pp. 72–86.

Taylor, M. (1999) Unwrapping stock transfers: applying discourse analysis to landlord communication strategies. Urban Studies Vol. 36 No. 1 pp. 121–135.

Taylor, M. (2000) Communities in the lead: power, organisational capacity and social capital. Urban Studies Vol. 37 No. 5–6 pp. 1019–1035.

Taylor, M. (2003) Neighbourhood governance: Holy Grail or poisoned chalice? Local Economy Vol. 18 No. 3 pp. 190–195.

Taylor, M. (2007) Community participation in the real world: opportunities and pitfalls in new governance spaces. Urban Studies Vol. 44 No. 2 pp. 297–317.

Tendler, J. (1997) Good Government in the Tropics. Baltimore: Johns Hopkins University Press.

Terrazas, A., Batalova, J. and Fan, V. (2007) Frequently Requested Statistics on Immigrants in the United States. Washington, DC: Migration Policy Institute.

Thompson, N. (2005) Inter-institutional relations in the governance of England's national parks: a governmentality perspective. Journal of Rural Studies Vol. 21 No. 3 pp. 323–334.

Tonkiss, F. and Passey, A. (1999) Trust, confidence and voluntary organisations: between values and institutions. Sociology Vol. 33 No. 2 pp. 257–274.

Tönnies, F. (1955) Community and Association (Gemeinschaft und Gesellschaft). London: Routledge & Kegan Paul.

United Nations Environment Programme (UNEP) (2007) Global Environmental Outlook 4. Malta: Progress Press Ltd.

von Braun, J. (2008) Rising food prices: what should be done? EuroChoices Vol. 7 No. 2 pp. 30–37.

Wacquant, L. (2003) Ethnografeast. A progress report on the practice and promise of ethnography. Ethnography Vol. 4 No. 1 pp. 5–14.

Wagenaar, H. and Cook, S.D.N. (2003) Understanding policy practices: action, dialectic and deliberation in policy analysis. In Hajer, M.J. and Wagenaar, H. (eds) Deliberative Policy Analysis. Understanding Governance in the Network Society, Cambridge: Cambridge University Press, pp. 139–171.

Wall, E., Ferrazzi, G. and Schryer, F. (1998) Getting the goods on social capital. Rural Sociology Vol. 63 No. 2 pp. 300–322.

Ward, N. and Lowe, P. (2004) Europeanising rural development? Implementing the CAP's second pillar in England. International Planning Studies Vol. 9 No. 2–3 pp. 121–137.

Ward, N. Lowe, P. and Bridges, T. (2003) Rural and regional developments: the role of the regional development agencies in England. Regional Studies Vol. 37 No. 2 pp. 201–214.

Ward, N. and McNicholas, K. (1998) Reconfiguring rural development in the UK: Objective 5b and the new rural governance. Journal of Rural Studies Vol. 14. No. 1 pp. 27–39.

Weber, M. (1947) The Theory of Social and Economic Organisation. Chicago: The Free Press.

Weber, M. (1968) Max Weber on law in economy and society (Rheinstein, M., ed; Shils, E. and Rheinstein, M., trans). New York: Simon and Schuster.

Wenger, E. (1999) Communities of Practice: Learning, Meaning and Identity. Cambridge: Cambridge University Press.

Williams, C.C. (2003) Developing community involvement: contrasting local and regional participatory cultures in Britain and their implications for policy. Regional Studies Vol. 37 No. 5 pp. 531–541.

Wilson, G.A. (2007) Multi-Functional Agriculture: A Transitional Perspective. Cambridge, MA: CABI International.

Woods, M. (1998a) Advocating rurality? The repositioning of rural local government. Journal of Rural Studies Vol. 14 No. 1 pp. 13–26.

Woods, M. (1998b) Rethinking elites: networks, space, and local politics. Environment and Planning A Vol. 30 No. 12 pp. 2101–2119.

Woolcock, M. (1998) Social capital and economic development: towards a theoretical synthesis and policy framework. Theory and Society Vol. 27 No. 2 pp. 151–208.

Woolcock, M. (2001) The place of social capital in understanding social and economic outcomes. Canadian Journal of Policy Research Vol. 2 No. 1 pp. 1–17.

Woolcock, M. and Narayan, D. (2000) Social capital: implications for development theory, research and policy. World Bank Research Observer Vol. 15 No. 2 pp. 225–249.

World Bank (1997) Participation Source Book. Washington, DC: The World Bank.

World Bank (2004) Local development discussion paper. Prepared for the International Conference on Local Development, Washington, DC: World Bank, 16–18 June, http://www1.worldbank.org/sp/ldconference/Materials/LDDPFinal.pdf, last accessed 31.07.08

Zetter, R., Griffiths, D., Sigona, N., Flynn, D., Pasha, T. and Beynon, R. (2006) Immigration, social cohesion and social capital. What are the links? York: Joseph Rowntree Foundation.

Index

Giddens, A. 38, 39, 47, 51, 67, 96
globalisation 8, 9, 12, 16; food policy
and 8–9
Goffman, E. 37, 40, 85
Goodwin, M. 9, 20, 60, 95, 100
governance (*see also* partnership)
9–24, 100–3, 118, 128, 133,
136 139–41; multi-scalar 7, 8,
12–15, 94; rural 3, 5–6, 20–5,
60, 137, 139–141
government 7–14, 96, 136, 137, 138;
decentralisation of 7, 8–10, ,
13–15, 20; and social capital 73
Granovetter, M. 69
group processes 61–3, 64, 68, 90, 92,
93
group relations 76–7, 84, 87, 90, 91, 92

H
Haiti, food riots 16
Hajer, M.J. 2, 96
Hanifan, 67
Harvey, D.R. 19
Hastings, A. 94, 97–9, 117
Hayward, C.R. 52–5
house prices 31
housing 27–32, 34, 36, 38, 48, 49, 63,
81–3, 87, 97, 108–12, 127, 130,
134–5
housing associations 20, 27, 29, 87;
regeneration activities 20, 26,
28–9, 33, 101
Housing Corporation 27, 28
Hunter, F. 49

H
ideology, differences of 22, 46, 62,
120–1, 128
IMF (International Monetary Fund) 17
India 15, 16
innovation 23, 28, 35, 141
Innovation and Good Practice scheme
(Housing Corporation) 28
integration, of rural development 3, 5,
8, 14, 15, 19–25
social interaction 11, 47, 54–68, 90–2,
115, 140; face-to-face 59, 61,
65, 77
interests (*see also* real interests): of
community 80, 83, 119, 136,
138; personal 81, 84, 103, 114;
and power relationships 56, 57;
professional 36–7, 43
international NGOs 19–20

involvement of community 14, 24, 98,
117, 121
Isaac, J.C. 46, 51, 55, 56, 137

J
Jessop, 9, 96
Jones, P. 105, 108, 132
Kilpatrick, S. 64, 71

L
land use 6
Latour, B. 42
LEADER initiative 8, 20–4
legitimacy 11, 24, 40, 43, 50, 53, 64,
67, 75, 132, 137; of community
power 24, 43–7
liberalisation 18, 22
Local Action Groups 23, 24
local government 7, 9–12, 33, 35, 48,
50, 51, 80, 135; modernisation
7, 12; reform of 7, 9
local strategic partnerships 11, 109,
Locke, J. 10, 44
Lowe, P. 8, 19, 21, 23
Lowndes, V. 9–12, 73, 88, 98, 118, 125
Lowry, K. 3, 61, 78
Lukes, S. 3, 42–49, 79, 84, 103, 105,
114, 127, 138, 140

M
Mandelson, Peter 16
manipulation, of decision-making
49–50, 79, 92, 127
Mann, M. 41, 48, 102, 120
marginalisation 35, 52, 98, 117, 131,
136–7
Marsden, T. 22
Marx, Karl 44, 45, 50, 67, 95, 100,
144
Mayer, M. 72, 73
McAreavey, R. 3, 39, 43, 55, 60, 143
meetings 29, 44, 49, 59, 60, 76–8, 80,
92, 109, 113, 115, 124, 127,
129, 135; attendance at 78, 105,
113; language use at 60, 83, 85;
location of 63, 127–8; micro-
politics of 61, 63, 75, 89, 92
Mexico, Strategic Community Centres
16, 19
micro-politics 4, 59–60, 140; in case
study 75–89; communication
and 3, 4, 75, 85–6, 88, 90–1;
language and 85–6, 90; legitimacy and 61, 64, 87–9, 91, 92;

Printed in the United States
by Baker & Taylor Publisher Services